Shrinking Cities:
Effects on Urban Ecology and Challenges for Urban Development

Marcel Langner / Wilfried Endlicher (eds.)

Shrinking Cities: Effects on Urban Ecology and Challenges for Urban Development

Second, revised edition

PL ACADEMIC RESEARCH

Bibliographic Information published by the Deutsche Nationalbibliothek
The Deutsche Nationalbibliothek lists this publication
in the Deutsche Nationalbibliografie; detailed bibliographic
data is available in the internet at http://dnb.d-nb.de.

Library of Congress Cataloging-in-Publication Data

Shrinking cities : effects on urban ecology and challenges for urban
development / Marcel Langner, Wilfried Endlicher, eds. —Second, revised
edition.
pages cm
ISBN 978-3-631-64294-8
1. Shrinking cities—Congresses. 2. Urban renewal—Germany—Congresses.
3. Urban renewal—Congresses. 4. Urban policy—Germany—Congresses.
5. Urban ecology (Sociology) —Germany—Congresses.
I. Langner, Marcel, 1973– II. Endlicher, Wilfried.
HT178.G4S56 2014
307.76—dc23

2013045106

ISBN 978-3-631-64294-8 (Print)
E-ISBN 978-3-653-04004-3 (E-Book)
DOI 10.3726/978-3-653-04004-3

Internationaler Verlag der Wissenschaften
Frankfurt am Main 2007
Second, revised edition 2014

PL Academic Research is an Imprint of Peter Lang GmbH.

Peter Lang – Frankfurt am Main · Bern · Bruxelles · New York ·
Oxford · Warszawa · Wien

www.peterlang.com

PREFACE TO THE SECOND EDITION

The editors are very glad to publish a second edition of this book with studies on urban ecology in shrinking cities. The book was one outcome of the second phase of the *Research Training Programme (Graduiertenkolleg) 780 "Perspectives of Urban Ecology"* (RTG 780), funded by the German Research Foundation. The third phase has ended now and most of the doctoral students have finished their theses, covering a wide range of topics from the urban geosphere, biosphere and anthroposphere. This interdisciplinary approach is important for studies in the urban environment and is also reflected in the contributions of this book. We made only minor changes to some contributions and updated the authors' address list.

Berlin, spring 2014

Wilfried Endlicher *Marcel Langner*

PREFACE TO FIRST EDITION

Berlin has been a focus of research into urban ecology for several decades. After the Second World War it was difficult to leave the 'political island' of Western Berlin. Academic personalities like Reinhard Bornkamm, Manfred Renger and Herbert Sukopp made a virtue out of necessity and dedicated their ecological research to this city, founding a scientific tradition which may today be called the 'Berlin School of Urban Ecology'. Basic textbooks like "Stadtökologie – das Beispiel Berlin" (ed. Herbert Sukopp) and "Stadtökologie" (eds. Herbert Sukopp and Rüdiger Wittig) summarised this research. Research in Eastern Berlin followed a similar, but less rigourous, direction, such as "Stadtklima Berlin" by Peter Hupfer and Frank-Michael Chmielewski.

It is on this solid basis that the *Research Training Programme (Graduiertenkolleg) 780 "Perspectives of Urban Ecology"* (RTG 780), funded by the German Research Foundation, began in 2002. After the fall of the Berlin Wall new challenges for urban ecological research arose in the unified city, the largest of Germany: Old factories closed, public utilities like slaughter houses became unnecessary and railroad yards were abandoned. The second phase of RTG 780 focused its integrated research on this subject.

The idea for this publication emerged in September 2006, when an international conference on urban ecology was organised by RTG 780 in Berlin. The book features seven talks given at the conference, two lectures held during the study programme of RTG 780 and two introductory chapters written by members of RTG 780.

The book begins with an overview of definitions and research concepts in urban ecology written by faculty members of RTG 780. The second chapter outlines the phenomenon of shrinking cities and its implications for urban ecology, with particular focus on brownfield sites. This second chapter was written by the programme's postgraduate students.

The two introductory chapters are followed by two papers concerning botanical aspects of shrinking cities. The first paper examines typical environmental factors that affect plants on brownfields and how plants adapt to them. The second paper discusses the influence of urban form on biodiversity in 21 German cities and the consequences for shrinking cities.

The next paper defines the connection between the natural system and its perception by humans in cities. The relationship between society and nature and how this 'socio-nature' is affected by shrinkage is examined.

The following couple of papers deal with aspects of urban planning in shrinking cities. The first paper is an overview of different approaches to urban regeneration and restructuring in Germany, and their effects on sustainability. The next paper focuses on the most important federal programme to deal with shrinkage in Eastern Germany, called *Stadtumbau Ost* ('Urban Restructuring East'). Another three papers discuss various aspects and several regional examples of shrinkage. Shrinkage leads to declining densities. Therefore, one paper discusses density as a tool to guide public works dealing with urban shrinkage. The next paper gives an example from Leipzig-Grünau regarding opportunities and challenges for urban green space planning in the context of shrinking cities. The last paper in this group examines how tram lines can be introduced to aid regeneration of brownfield sites and gives examples from Manchester, Sheffield and Amsterdam.

The book concludes with a reflection on urban shrinkage during the life cycles of city archetypes.

The editors wish to thank the twelve professors of the three Berlin Universities that combined their efforts at the end of the last decade to realize RTG 780. We are grateful to the postgraduate students of the first two phases of the programme who discovered unprecedented results in their research and collated new knowledge about the nature, environment, and natural and human dimensions of the European metropolis Berlin. We would like to express our thanks to the members of the Integrated Graduate Education and Research Training Group on Urban Ecology of the University of Washington in Seattle, who regularly discussed characteristic processes of shrinking (and growing) cities with us. We wish to thank Rachael Marshall for useful linguistic comments and Matthias Grätz for his invaluable technical help. Finally we would like to thank the German Research Foundation for funding this volume.

We are looking forward to a fruitful continuation of the scientific work on urban ecology in Berlin in general and of the RTG 780 in particular.

Berlin, summer 2007

Wilfried Endlicher *Marcel Langner*

CONTENTS

URBAN ECOLOGY – DEFINITIONS AND CONCEPTS

Wilfried Endlicher, Marcel Langner, Markus Hesse, Harald A. Mieg, Ingo Kowarik, Patrick Hostert, Elmar Kulke, Gunnar Nützmann, Marlies Schulz, Elke van der Meer, Gerd Wessolek, Claudia Wiegand

1. Introduction

Earth's population more than doubled during the second half of the twentieth century: from approximately 2.5 billion in 1950 to over 6 billion in 2000, and at the time of writing in 2007 has reached a figure of over 6.6 billion. Alongside this exponential growth of population is another important demographic trend: According to the United Nations, the anticipated population growth between 2000 and 2030, approximately 2 billion people, will be concentrated in urban areas (UN 2004). The 21st century will be the century of urbanisation. By the year 2030 more than 60 per cent (4.9 billion) of the estimated world population (8.1 billion) will live in urban settlements, compared to 29 per cent in 1950. The 50 per cent mark is expected to be reached in the year 2007. In 2025, more than a dozen urban agglomerations will have over 20 million inhabitants, and some will have over 30 million. 23 of the 25 biggest urban agglomerations on the planet will be in Africa, Asia, and Latin America, rather than in Europe or North America (Kraas 2003). These megacities are considered 'hotspots' of global change (Kraas 2007).
Urbanised areas cover between approximately one and six per cent of Earth's surface, yet they have extraordinarily large ecological 'footprints' and complex, powerful, and often indirect effects on ecosystems (Rees & Wackernagel 1994).

2. Ways to define urban ecology

The aim of 'Urban Ecology' is to study these effects. According to Sukopp & Wittig (1998), the term 'Urban Ecology' (in German *Stadtökologie*) can be defined in two ways. Within the natural sciences, urban ecology addresses biological patterns and associated environmental processes in urban areas, as a subdiscipline of biology and ecology. In this sense, urban ecology endeavours to analyse the relationships between plant and animal populations and their communities as well as their relationships to environmental factors including human influences. From this perspective, the research is unconstrained by anthropocentric evaluations. However the second, complementary, definition implies the anthropocentric perspective. Here, urban ecology is understood as a multidisciplinary approach to improving living conditions for the human

population in cities, referring to the ecological functions of urban habitats or ecosystems for people – and thus including aspects of social, especially planning, sciences. From an even broader view, cities can be considered as emergent phenomena of local-scale, dynamic interactions among socio-economic and biophysical forces. These are both complex ecological entities that have their own unique internal rules of behaviour, growth, and evolution, and important global ecological forcing influences (Alberti et al. 2003). Urban ecology is the study of ecosystems that includes humans living in cities and urbanising landscapes. It investigates ecosystem services which are closely linked to patterns of urban development (Alberti 2005).
Urban ecology is an interdisciplinary field that supports societies' attempts to become more sustainable. It has deep roots in many disciplines including geography, sociology, urban planning, landscape architecture, engineering, economics, anthropology, climatology, public health, and ecology. Because of its interdisciplinary nature and unique focus on humans and natural systems within urbanised areas, 'urban ecology' has been used variously to describe the study of humans in cities, nature in cities, and the coupled relationships of humans and nature (Marzluff et al. 2008; Fig. 1).

3. Conceptual history of research in urban ecology

Urban ecology has many disciplinary roots. In recent decades, the conceptual approach of the 'Berlin School of Urban Ecology', promoted mainly by Herbert Sukopp since the 1970s, was influential (Wächter 2003). By this approach, urban habitats and associated environmental processes were analysed at local and regional scales by different disciplines of natural sciences. This includes biodiversity patterns as well as characteristics of urban soils and climate and their variation in time and space due to changing urban land uses. While the contemporary ecosystem approach of Duvigneaud (1974) mainly addressed fluxes of energy and matters at the city level, the Berlin approach focussed on the explicit spatial variation of ecological components within urban environments. This also led to the first model of a city characterised by idealised variation in climate, soils, terrain, vegetation and fauna along a transect from the densely built-up city centre to the outskirts (Sukopp 1973; Fig. 2). Sukopp distinguishes a core surrounded by three rings – the densely built-up central core area, a ring with more open space, where some smaller cores of densely built-up subcentres may be found and finally the interior and exterior border zones. Concentric models of the spatial organisation of land-uses have existed since von Thünen (1826) and Burgess (a member of the Chicago School of Social Ecology; Burgess 1925 and 1929) developed a model of the concentric structure of cities from the perspective of social sciences. Sukopp was the first to qualify such a model with a broad array of ecological factors.

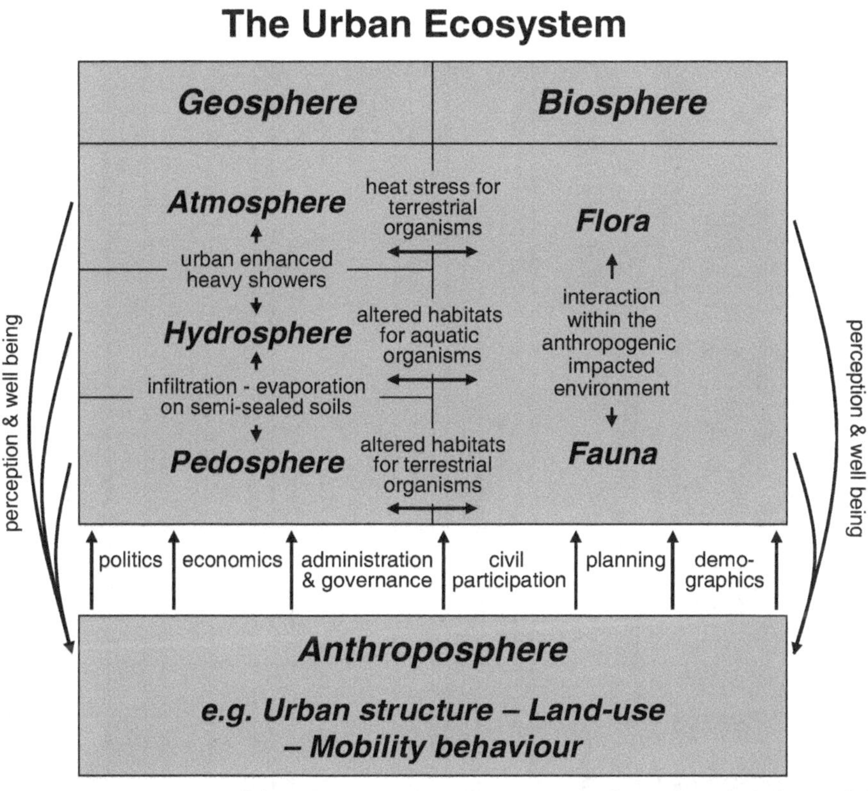

Fig. 1: Basic components of the urban ecosystem; this concept is focused on the spheres of the Earth system which are important for cities. The processes between different spheres and the impacts of the anthroposphere (six selected examples) are of special interest (MARZLUFF et al. 2008; modified)

Perhaps the most often reproduced diagram in urban ecology shows a transect through the concentric rings and its consequences for climate, soil and water, topography, vegetation and animal life in the different urban zones (Fig. 2).

Many studies of urban ecology follow Sukopp's transect approach, comparing the specific ecological situation of each zone with the others, and the whole city with its surrounding environment (e.g. the *urban heat island* characterises the maximum temperature differentiation between a climate station in the core area and another outside the built-up area; ALCOFORADO & ANDRADE 2007). The urban heat island of densely built-up environments is an important factor of additional heat stress in summer months. Lower work efficiency, enhanced morbidity and cardio-vascular dis-

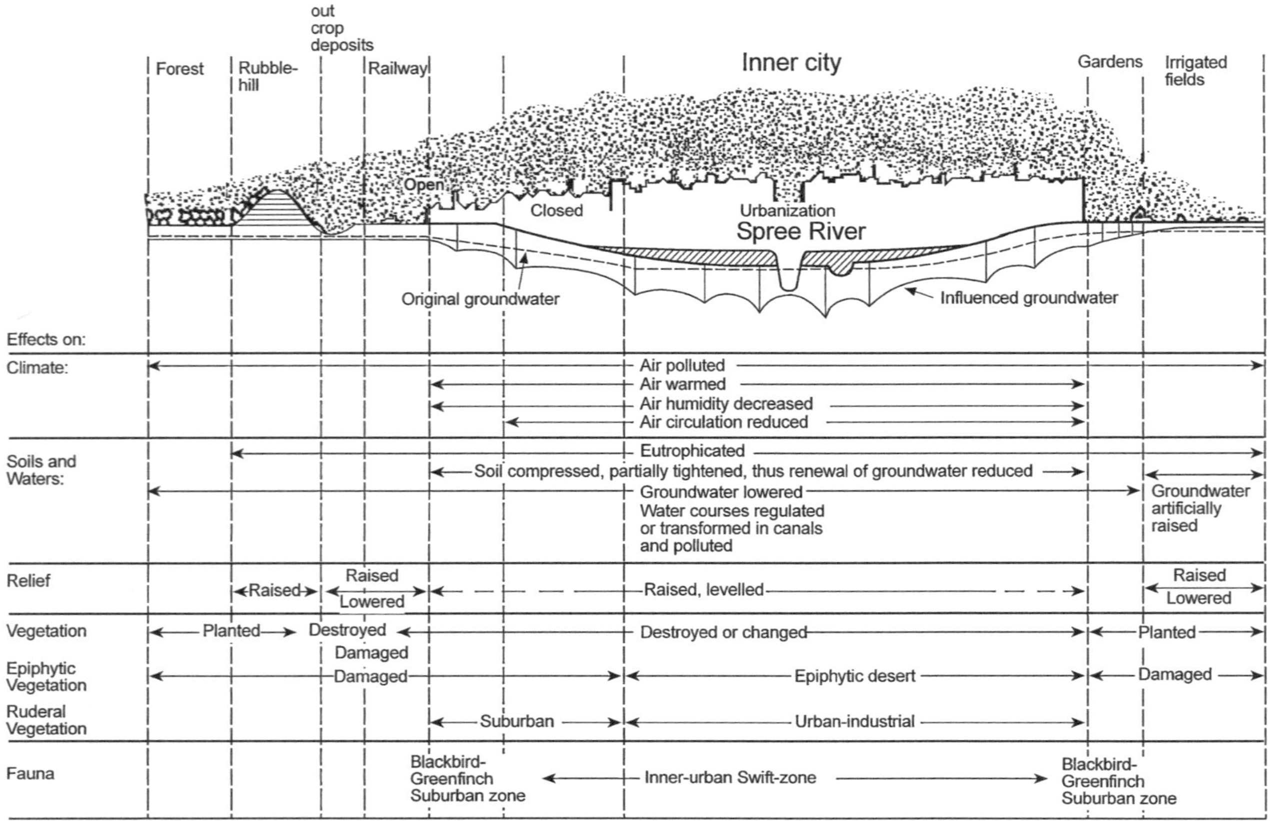

Fig. 2: Transect through the urban built-up structure of Berlin and the ecological consequences for different spheres; this classic concept concentrates on the impacts of urbanisation for five layers: climate, soil and water, relief, vegetation and fauna (adapted after SUKOPP 1973)

eases are related to high solar radiation, air temperatures and humidity (KOVATS & JENDRITZKY 2006).

The traditional model of the multinuclear city proposed by HARRIS & ULLMAN (1945) from the Chicago School of Social Ecology is another approach to urban ecology. This includes the classification of built-up structures of cities (*Stadtstrukturtypen* or *Baukörperstrukturen* in German). The ecological conditions of each structural type (e.g. industrial area, central business district, suburb with housing function, middle class housing quarter) are investigated and their characteristics may be compared. WITTIG, SUKOPP & KLAUSNITZER (1998) gave a detailed description of the built-up types in German cities. WICKOP et al. (1999) used this model for their ecological studies of Leipzig. This is another widely used method in urban ecology.

Urban ecology can be understood as a spatial science in the same way as geography. Therefore the scale of the studies to be carried out is important. Three different scales should be distinguished, especially in larger cities: the micro-scale of the local neighbourhood with its special built-up characteristics where the study or field experiment is carried out, the meso-scale of the district, which features a combination of different land use (built-up) types and finally the macro-scale of the total urban area, sometimes composed of different administrative entities or even cities. The results of the studies may permit a certain generalisation for the three scales and some typical neighbourhoods/districts/cities may be identified, leading to prototypes of a 'virtual city' (Fig. 3).

Urban ecology addresses processes in space and time. Besides the spatial dimension, four main processes of change are the focus of recent research: changes in urban biodiversity, climate, human demography and economy:

Urban land use significantly affects **biodiversity patterns**. Until the 1960s cities were perceived as 'biological deserts', whereas they are currently considered as 'hotspots' of botanic and animal diversity. Species respond quite differently to urbanisation, with a decline in native species and increase in introduced species as a general trend (KOWARIK 1990). These changes in urban biota are currently regarded as major drivers of global homogenisation (MCKINNEY 2006). However, regional studies have demonstrated that both native and non-native species richness is higher in urban areas than adjacent areas and that non-native species may also contribute to the dissimilarity of urban floras (KÜHN & KLOTZ 2006). Future analysis should thus examine the role of cities in endangering or conserving biodiversity in depth.

Cities are important drivers of **climate change** because about 75 per cent of greenhouse gas emissions are produced in urban territories. Simultaneously, however, cities are especially vulnerable to climate change as Working Group II of the Intergovernmental Panel of Climate Change concludes in its 4th Assessment Report (IPCC 2007). Many components and processes of the earth system are affected, especially the atmosphere (rising temperatures and extreme weather events), the hydrosphere

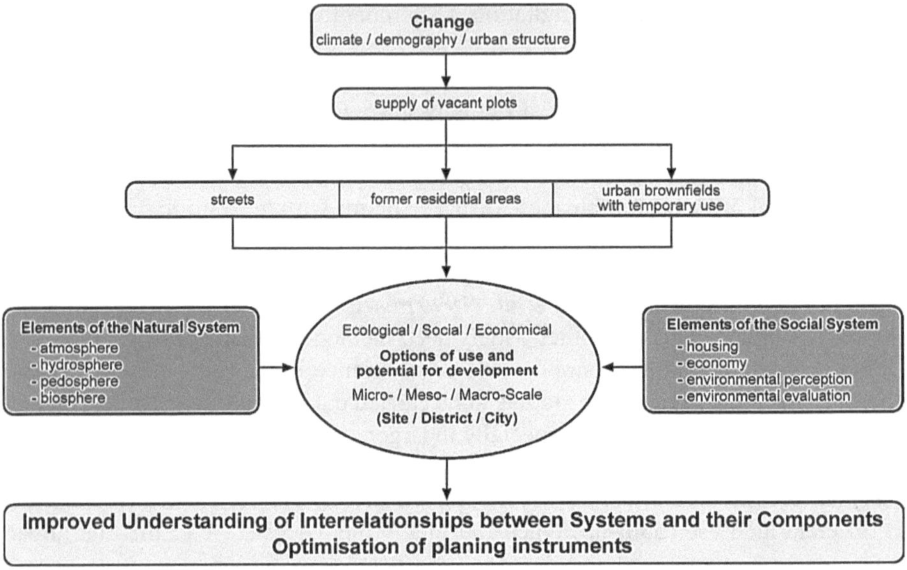

Fig. 3: Example of an approach in urban ecology that considers three scales, with specific reference to processes of change

(rising water levels) and the biosphere (drastic changes to biodiversity). Urbanised areas may serve as subject of field experiments in order to investigate plant responses to climate change, since temperature and CO_2-concentrations are already increased in cities (ZISKA et al. 2003). Impacts are highly variable, but include an increased burden of diseases, increased morbidity and mortality from more frequent and intense heat waves superimposed on the urban heat island. Coastal megacities are particularly at risk from floods, storms and droughts (KRAAS 2003 & 2007).

Demographic change may also exert an influence on the anthroposphere. In highly industrialised countries people are growing older than ever before, while birth rate is simultaneously decreasing in countries like Germany (KAUFMANN 2005). The proportion of senior citizens is expected to increase, and the pyramid of population is likely to change its shape. This causes modifications of behaviour and demand for living space, for example. However, it is indicative that demographic changes offer potential for improving the ecological conditions of cities, not only due to a reduced number of individuals and therefore demand for water, energy, transport etc., but also in the context of a decreasing pressure on land use and the possibility of alternatives to the classical growth of urban development. Conversely, to ensure cost-efficient technical infrastructures, building density should not fall below a certain threshold (see paper of WESTPHAL in this book).

Economic Change is one of the most important factors for the function and development of urban agglomerations. A town's role in the interregional and supranational network of cities is affected by its economic structure and, in addition, existing economic activity dominates the urban environment. Cities have undergone rapid changes to their economic structures during recent decades: They are becoming increasingly integrated into global supply and demand systems which depends on the process of globalisation. Alongside these developments a key factor in advanced economies for urban agglomerations is the switch from industrial to service based economies; spatial characteristics of this change are the appearance of brownfields on former industrial land and growing demand for spaces for high-ranking services. The four above-mentioned changes are important issues to be taken into account in future urban ecological research and planning processes (STONE 2005).

Urban ecology is an interdisciplinary science where elements of the natural spheres and the anthroposphere with its socio-economic aspects must be taken into account. Therefore, integrated approaches are necessary for a more comprehensive understanding of the ongoing processes. Research clusters, studying a particular question from different disciplinary perspectives, may be especially useful in urban ecology. Clusters may include elements of the abiotic spheres (atmosphere, hydrosphere and pedosphere) and the biotic sphere (flora and fauna), which together form 'the natural system of a city', or the anthroposphere (society and economy), which forms 'the socioeconomic system of a city' (Fig. 1 and 3).

4. Current research of the 'Urban Ecology' Research Training Group

Five Berlin universities and research institutes founded the Research Training Group (*Graduiertenkolleg*, RTG 780) 'Urban Ecology'[1] on April 1st 2002. The programme will end in 2011 and is divided into three terms, with each term lasting three years. The research approaches of the three terms serve as examples of the general concepts mentioned above (Tab. 1).

During the first term RTG 780 organized its research on a northwest-southeast transect through the Berlin Metropolitan Area, from the inner core to the outskirts. Research on the growing conditions of neophytes, such as local temperature (VON DER LIPPE, SÄUMEL & KOWARIK 2005), urban air quality (WOLF-BENNING, DRAHEIM & ENDLICHER 2005), soil conditions in urban environments (NEHLS et al. 2006), exchange of species along urban-rural gradients (VON DER LIPPE & KOWARIK 2008) and the habitat of the kestrel (KÜBLER & ZELLER 2005) present the results of such

1 Sponsored by the German Research Foundation (DFG)

Tab. 1: Main research concepts of the three terms of RTG 780

	Duration	Main research concept
Term 1	2002 – 2005	Classical transect approach
Term 2	2005 – 2008	Investigation of brownfield sites as unique urban structures
Term 3	2008 – 2011	Research clusters as an integrated approach

studies along an urban transect. Furthermore, Tobia Lakes and Sonja Pobloth (2005) investigated urban habitat networks and Marit Rosol (2005) carried out research on community gardens.

However, specific subjects such as urban brownfields have been selected to be investigated in detail, for various reasons. This was the case for the second term of the RTG 780. Urban brownfields represent a typical feature in shrinking cities and are of special interest to urban ecology; they may be perceived in a variety of ways. They offer – perhaps only for a limited time span – habitats for plants and animals, and they play a role in the urban economy. Streets and roadsides are another important urban structures, sometimes occupied by trees, bushes and private gardens. Surface waters in the form of watercourses or lakes contribute to the appearance of a city; they influence the microclimate of their surroundings and provide habitats for aquatic plants and animals. Roadsides and brownfields, as well as urban parks and surface waters offer space for urban nature and must be taken into account as elements of the quality of urban life. The quality, defined by specific size, composition and environmental health of the neighbourhood's nature plays an important role for the well-being of the residents and public health in general.

Many problems in cities cannot be understood if natural and social sciences act separately. Urban ecology cannot be considered as simply a subfield of bioecology, but must be integrated with the human dimension at all levels and scales, in the formulation of research topics as well as the assembly of research teams. For this purpose, various areas of expertise connected to urban ecology are combined, ranging from remote sensing via surveys of local field conditions for the sustainable establishment of neophyte or bird populations inside the city, to molecular tools for assessing environmental health. This is combined with an integrated view covering sociological and economical driving forces on city developments including their consequences for the perception and well being of the citizens themselves. The following examples of research topics and the integrated approach developed by the RTG 780 for its third term demonstrate this more explicitly.

Cluster 1: Biodiversity and optimising ecological functions of roadside areas
Roadsides characterised by high levels of physical stress and pollution are ubiquitous habitats in urban environments. They may function as habitats for plant and animal species, dependent on the species-specific sensitivity to physical stress, pollution and maintenance and associated changes in soils and thermal conditions. The emergence of roadside vegetation may conflict with public interests in safety or tidiness but provides important ecological services such as mitigation of temperature increase, sequestration of harmful substances, or habitat functions. Similarly, urban watercourses are anthropogenically affected by constructional changes, such as sheet pile walls which together with siltation and reasonable high pollution load shift abundancies to more tolerant species. A better understanding of ecological mechanisms and functions as well as of the public perception of different habitat types will support optimised strategies to develop and maintain areas alongside urban roadsides and watercourses.

Cluster 2: Re-use of former housing estates
All highly industrialised countries are currently facing the problem of deindustrialisation, particularly in the sectors of heavy industries. Examples are large areas in Eastern Germany and the Ruhr Area in Western Germany, parts of the British Midlands and of the United States of America, e.g. Detroit or Pittsburgh in the Northeast. Large areas that had formerly been used by heavy industries are now vacant.
Many of these regions also suffer from a decrease in population due to migration and demographic change, or as an immediate response to job losses. Therefore, the process of shrinkage is not exclusive to old industrial estates, but also occurs on sites used for technical infrastructure, services and housing. The latter are also called *Wohnfolgelandschaften* (former housing estates) in German. This term refers to settlement areas characterised by large vacant plots caused by the demolition of buildings on the edges of, or even in the centres of, cities.
These changes create challenges for further urban development strategies. Risks and opportunities will be clarified by combined approaches to socio-economic, environmental psychological and ecological research. Strategies to encourage biodiversity and the public perception of these are tested using an experimental approach. The interdisciplinary approach intends to support planning strategies of decision makers.

Cluster 3: Strategies for temporarily used urban sites
Many urban sites whose original functions are defunct (e.g. marshalling yards or former housing estates) can now be re-used for other purposes. These do not necessarily have to be permanent. Temporarily used urban sites serve as flexible instruments for urban planning and development. This planning tool is used to avoid temporary difficulties arising from property conditions.

It is assumed that brownfields and other temporarily used urban sites are just as important for the urban ecosystem as permanently used areas like urban parks or forests. A key consideration is the availability of temporarily used urban sites of different sizes and ecological impacts. This availability is governed by three factors: (1) the urban land market, (2) political and planning regulations and (3) processes of opinion forming and decision making. These three socio-economic and political/planning categories are also the most important factors in the system of land use in urban agglomerations. Temporarily used urban sites are investigated in the context of these categories to improve their impact on the urban ecosystem.

Cluster 4: Psychological health and state of city residents

Urban ecosystems are subject to short-term changes resulting in objective modifications to the environment. To understand the reaction of residents to these changes it is important to know whether objective modifications lead to a change in subjective perceptions and assessments. Behavioural decisions of city residents may then be influenced by these objective modifications.

Both objective modifications of the urban environment and changes to subjective perceptions induce specific sensitivities in residents. For example, changes to surface water and groundwater quality and structure of urban littoral zones can result in health risks.

Two research stages are preferred: Firstly, the determination of objective measurable parameters which are thought to be particular harmful in urban environments, such as heat stress or water scarcity. Modelled scenarios based on measured data are helpful to illustrate possible urban environmental developments. Secondly, the analysis of perception and effective assessment by residents must be studied.

5. Challenges for urban ecology and the city

It has emerged that urban locations can be ecologically abundant due to the specific characteristics of each subsystem and their densities. Such systems demonstrate higher biodiversity than some of the areas traditionally perceived as near-natural, e.g. agricultural areas.

Amin and Thrift (2002) emphasised that nature and the city can no longer be considered dichotomic, as a matter of contradiction per se, yet the boundaries between them have blurred to a significant extent. This is evident in the diversity of species hosted by modern cities. In turn, this may prove to alter human perception of nature. As Amin and Thrift stated, the Environmental Agency of London sold about 200,000 fishing-rod licences to people in London in 1999 – almost a third of the annual total for England and Wales (Amin & Thrift 2002:44). Accordingly, the recently noted appearance of wild animals in cities (e.g. foxes, wild pigs and mountain lions) and the increasing invasion of non-native plants indicate the changes in urban and thus societal relations to nature.

The discovery of the ecological value of man-made and settled areas is conntected to earlier works about the European *Kulturlandschaft* (cultural landscape), the positive perception of which was precisely due to the impact of human activity, and not despite it (cp. Ellenberg 1982). The emergence of urban ecosystems represents a new phase in the line of transfer between pristine, or natural, to culturally-shaped ecosystems, as conceptualized by Kowarik's model of 'four natures' which contrasts the urban-industrial nature as 'fourth nature' with preceding stages of natural and cultural landscapes (Kowarik 2005).

It is widely accepted that urbanisation has a significant effect on existing rural and natural landscapes, but the perception of urban-industrial habitats has yet to progress from one-sided negative evaluation. They may contribute to the stabilisation and improvement of natural living conditions, regardless of whether they are in regular use. Urban ecology studies have re-introduced the significance of urbanised areas as ecologically relevant. In this context, conceptualisation of urban space as an ecological entity, as described above, represents a paradigm shift in environmental research.

A novel concept of urban ecology arises from the discussion of these and various American concepts, with particular focus on the human dimension. Or, as Alberti et al. (2003) have suggested, the actual challenge is "integrating humans into ecology". There is a profound disconnection between nature and wilderness on one hand, and the built-up environment of cities on the other:

Cities are usually so large that city dwellers' contact with nature is difficult and often it is only poor industrial agriculture or tree monoculture that can be easily accessed ouside the city, whereas biodiversity inside the city is high and differentiated – but often not recognised. The ecologic and economic values of the 'fourth nature', or 'new urban wilderness', are not yet broadly appreciated. City dwellers spend most of their time indoors – in environments with artificially heated or cooled ambient air, treated drinking water from pipes, soil in flower pots with ornamental plants, and small pets. This 'home nature' – a fifth type of nature – goes some way to replacing the outdoor type that city dwellers are disconnected from. However, human well-being, work efficiency and health also depend on intact natural elements close to daily life in cities. The colourful, spotted 'harlequin pattern' (Sukopp) is not only typical for urban biotopes, but can be found in a multitude of local climates and soil sites, too. Human activity must be considered as an essential part of urban ecology, and the integration of geo-biosphere- and anthroposphere-approaches is urgently needed. Good practice examples of such integrated actions can be developed at ecological hotspots in cities (Fig. 4).

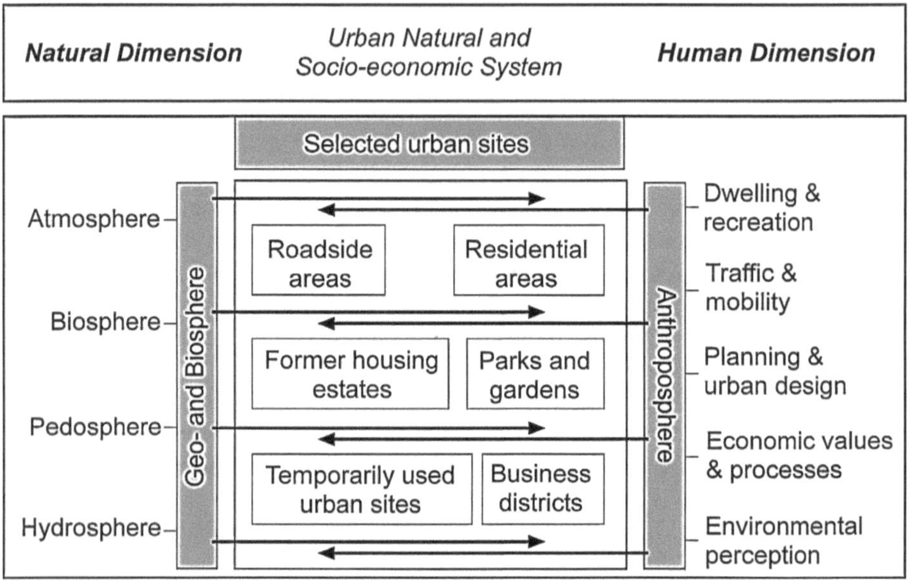

Fig. 4: Concept of integrated research in urban ecology: The main focus must be on the human dimension and its interferences with the urban natural system. A robust integration of human activities into urban ecology seems necessary due to the distinct disconnection between nature and daily urban life. To accomplish this, research should be carried out simultaneously from environmental and social sciences at the same urban sites

Given this more differentiated view of urbanised or industrialised areas from the urban ecology perspective, another basic question arises with regard to future urbanisation processes. As mentioned above, the year 2007 is considered to have witnessed a historical shift in human existence, where for the first time the majority of people live in cities rather than rural areas. Increasing land use and degradation by construction and settlement unquestionably create additional problems. However, it remains to be proven whether it would be more ecologically favourable to distribute a given urban population across a much larger area. The two extreme cases of urbanisation, the sprawling metropolitan areas of the Western world and the exploding megacities of developing nations, may easily be assessed as unsustainable. But what is the shape of a sustainable urban future? What are the dynamics and processes that would make it work?

Urbanisation can be understood as a process of spatial concentration supported by the economics of agglomeration and the rich socio-cultural amenities offered by cities. However, urban magnets are almost inevitably facing the risks and disadvantages of agglomeration, such as congestion, air quality problems, scarce housing supply etc. As a consequence, processes of decentralisation almost always accompany urban growth. How should the precise impact of these different, and to some extent competing, forces be assessed? Is there a model of urbanisation (undispersed, yet not too dense) that could be recommended as a solution? Clarification of this question may represent one of the future tasks of urban studies in general and urban ecology in particular.

Finally, urban ecology is relevant to urban policy, since any improvement to living conditions in urban areas requires a societal and individual awareness of the problem, as does the attempt to make cities more sustainable in terms of the natural environment. As Wolch (2007) has stated, the challenges for policy led by urban ecology are threefold: first, the city's ecological integrity must be reinstated, which means recreating a green matrix in them, in order to bring plants and animals back to where the majority of people live; second, the systems of production and consumption must be redesigned to address the global problem of an unsustainable 'metabolism'; third, urban citizenship must be revived, not only to make the ecological transition acceptable and accepted in terms of society, but also in order to pursue social and ecological justice.

References

Alberti, M., Marzluff, J.M., Shulenberger, E., Bradley, G., Ryan, C. & C. Zumbrunnen (2003): Integrating humans into ecology: Opportunities and challenges for studying urban ecosystems. Bioscience 53(12): 1-11.

Alberti, M. (2005): The effects of urban patterns on ecosystem function. International Regional Science Review. 28(2): 168–192.

Alcoforado, M.J. & H. Andrade (2007): Global warming and the urban heat island. In: Marzluff et al. (in press).

Amin, A. & N. Thrift (2002): Cities: Reimagining the urban. Polity Press. Cambridge.

Burgess, E.W. (1925): The growth of the city: Introduction to a research project. In: Park, R.E., Burgess, E.W. & McKenzie, R.D. (eds.): The City. University of Chicago Press. Chicago.

Burgess, E.W. (1929): Urban areas. In: Smith, T.V. & White, L.D. (eds.): Chicago: An Experiment in Social Science Research. University of Chicago Press. Chicago.

Duvigneaud, P. (1974). L'écosystème "Urbs." Mém. Soc. Roy. Belg. 6: 5–35.

Ellenberg, H. (1982): Vegetation Mitteleuropas mit den Alpen in ökologischer Sicht. Ulmer. Stuttgart.

Harris, C.D. & E.L. Ullmann (1945): The Nature of Cities. Annals of the American Academy for Political Science 242: 7-17.

IPCC (2007): Climate Change 2007: Impacts, Adaptation and Vulnerability. Working Group II Contribution to the Intergovernmental Panel on Climate Change Fourth Assessment Report. Summary for Policymakers. Online available at http://www.ipcc-wg2.org/index.html (25 June 2007).

Kaufmann, F.-X. (2005): Schrumpfende Gesellschaft. Vom Bevölkerungsrückgang und seinen Folgen. Bonn.

Kovats, S. & G. Jendritzky (2006): Heat-waves and human health. In: Menne, B. & K.L. Ebi (eds.): Climate change and adaptation strategies for human health. Darmstadt: 63-97.

Kowarik, I. (1990): Some responses of flora and vegetation to urbanization in Central Europe. In: Sukopp, H., Hejny, S. & I. Kowarik (eds.): Plants and plant communities in the urban environment. SPB Academic Publishing, The Hague: 45-74.

Kowarik, I. (2005): Wild urban woodlands: Towards a conceptual framework. Kowarik, I. & S. Körner (eds.): Wild Urban Woodlands. Berlin & Heidelberg: 1-32.

Kraas, F. (2003): Megacities as global risk areas. Petermanns Geographische Mitteilungen 147: 6-15.

Kraas, F. (2007): Megacities and global change: key priorities. Geographical Journal 173: 79-82.

Kübler, S. & U. Zeller (2005): The kestrel (*Falco tinnunculus* L.) in Berlin: Feeding ecology along an urban gradient. Die Erde 136: 153-164.

Kühn, I. & S. Klotz (2006): Urbanization and homogenization – Comparing the floras of urban and rural areas in Germany. Biological Conservation 127: 292-300.

Lakes, T. & S. Pobloth (2005): The evolution of approaches to create urban habitat networks – the case of Berlin. Die Erde 136: 179-200.

Marzluff, J.M., Shulenberger, E., Endlicher, W., Alberti, M., Bradley, G., Ryan, C., Simon, U. & C. ZumBrunnen (2008) (eds.): Urban Ecology. An international perspective on the interaction between humans and nature. New York.

McKinney, M. L. (2006): Urbanization as a major cause of biotic homogenization. Biological Conservation 127: 247-260.

Nehls, Th., Sokolowska, Z., Hajnos, M., Jozefaciuk, G. & G. Wessolek (2006): Pore-system characteristics of pavement seam materials of urban sites. Journal of plant nutrition and soil science 169: 16-24.

Rees, W. & Wackernagel, M. (1994): Ecological footprints and appropriated carrying capacity: Measuring the natural capital requirements of the human economy. In: Jansson A.-M., Hammer, M., Folke, C. & R. Costanza (eds.). Investing in natural capital: The ecological economics approach to sustainability. Washington: 362-390.

ROSOL, M. (2005): Community gardens – A potential for stagnating and shrinking cities? Examples from Berlin. Die Erde 136: 165-178.

STONE, B. Jr. (2005): Urban heat and air pollution. An emerging role for planners in the climate change debate. J.American Planning Association 71(1): 13-25.

SUKOPP, H. (1973): Die Großstadt als Gegenstand ökologischer Forschung. Schr.R. Ver. Verb. naturwiss. Kenntn. Wien 113: 90-140.

SUKOPP, H. & R. WITTIG (1998): Was ist Stadtökologie? In: SUKOPP, H. & R. WITTIG (eds.): Stadtökologie. 2nd ed. Stuttgart etc.: 1-12.

UN (2004): World urbanization prospects: The 2003 revision. United Nations – Department of Economic and Social Affairs – Population Division. New York. Online available at http://www.un.org/esa/population/publications/publications.htm (04 May 2007).

VON DER LIPPE, M., SÄUMEL, I. & I. KOWARIK (2005): Cities as drivers for biological invasions – The role of urban climate and traffic. Die Erde 136: 123-143.

VON DER LIPPE, M., I. KOWARIK (2008): Do cities export biodiversity? Traffic as dispersal vector across urban–rural gradients. Diversity and Distributions 14: 18-25.

VON THÜNEN, H.J. (1826): Der isolierte Staat in Beziehung auf Landwirtschaft und Nationalökonomie. Berlin. (Reprint Stuttgart 1966).

WÄCHTER, M. (2003): Die Stadt: umweltbelastendes System oder wertvoller Lebensraum? Zur Geschichte, Theorie und Praxis stadtökologischer Forschung in Deutschland. UFZ Ber. 9. Leipzig.

WICKOP, E., BÖHM, P., EITNER, K. & J. BREUSTE (1998): Qualitätszielkonzept für Stadtstrukturtypen am Beispiel der Stadt Leipzig. UFZ-Bericht 14, Leipzig/Halle.

WITTIG, R., SUKOPP, H. & B. KLAUSNITZER (1998): Die ökologische Gliederung der Stadt. In: SUKOPP, H. & R. WITTIG (eds.): Stadtökologie. Stuttgart etc.: 316-372.

WOLCH, J. (2007): Green urban worlds. Annals of the Association of American Geographers 97(2): 373-384.

WOLF-BENNING, U., DRAHEIM, T. & W. ENDLICHER (2005): Particulate matter and nitrogen dioxide in Berlin's air – spatial and temporal differences. Die Erde 136: 103-121.

ZISKA, L.H., GEBHARD, D.E., FRENZ, D.A., FAULKNER, S., SINGER, B.D., & J.G. STRAKA (2003): Cities as harbingers of climate change: common ragweed, urbanization, and public health. J. Allergy and Clinical Immunalogy 111 (2): 290-295.

SHRINKING CITIES – A NEW CHALLENGE FOR RESEARCH IN URBAN ECOLOGY

Miriam Fritsche, Marcel Langner, Hadia Köhler, Anke Ruckes, Daniela Schüler, Betka Zakirova, Katharina Appel, Valeska Contardo-Jara, Ellen Diermayer, Mathias Hofmann, Christoph Kulemeyer, Peter Meffert, Janneke Westermann

1. Introduction

The history of urban development is not a history of constant growth. It is characterised by phases of growth, stagnation and crises due to new economic, social and political conditions. Cities have been affected by shrinking processes throughout their development (Benke 2005). 'Shrinkage' and 'shrinking cities' are current phenomena of urban development in Germany and other regions worldwide. Causes, types and impacts of shrinking cities are studied, but there are few academic investigations dealing with the effects of shrinkage on urban ecology. Although processes of shrinkage and the consequent emerging brownfields could enable sustainable urban development, e.g liberate new urban spaces for city dwellers, flora and fauna, and hence improve the quality of natural resources (e.g. water, air, soil) and urban life. Therefore, this article discusses the principal question of how shrinkage affects urban ecology.[1]

The first section outlines the current state of the German research into shrinking cities as well as the impact of shrinkage on processes of urban development in Germany. The scope of the second section broadens to the global scale of urban shrinkage to demonstrate the spread of shrinking cities and the different reasons for it as a world-wide phenomenon. Subsequently, focussing on the local scale, the case of Berlin-Brandenburg will be discussed for two reasons: First, Berlin is an example of ongoing processes of simultaneous shrinkage and growth; furthermore the urban ecology of the city is well researched. The socio-scientific framework of shrinkage is followed by a section that selects the meaning of urban brownfields as a central theme for interdisciplinary research. The potential of brownfields for improving sustainable development will be exemplified with regard to various approaches in natural, environmental and social sciences.

---

1 The context of 'shrinkage' and 'urban ecology' is the main research topic for postgraduate students in environmental and social sciences in the interdisciplinary research training group (RTG 780) "Shrinking Cities: Structural Changes as Opportunities for the Development of Urban Nature and Enhanced Environmental Quality for City Dwellers"; sponsored by the German Research Foundation (DFG).

2. 'Shrinkage' and 'shrinking cities' as subjects in German research on urban development and urban politics

In general, 'shrinkage' means a decline of any quantity in terms of diminution, reduction or downsizing. In current German research, the term is used in two connected contexts: As a result of the analysis of demographic change ('shrinking society') and in studies on urban development and urban politics ('shrinking cities'). Demographers are expounding the problems of the foreseeable acceleration of Germany's population decline (KAUFMANN 2005). Urban research is dealing with causes, forms and consequences of urban decline following processes of de-industrialisation and decreasing birth rates (see various articles in: BERLINER DEBATTE INITIAL 2002, LANG & TENZ 2003, HANNEMANN 2004, KABISCH et al. 2004, HEINEBERG 2005, GLOCK 2006). No general definition of 'urban shrinkage' has been accepted in German research so far (BRANDSTETTER et al. 2005), but there is mutual consent with regard to its causes, namely de-industrialisation, suburbanisation and demographic change. The interrelation of these different processes produces a vicious circle in current urban development: *"Decrease in population is caused by migration; high unemployment is a result of decrease in jobs; decline of population and jobs lead to a decrease of purchasing power and deficiencies in public revenues. Less private and public financial resources cause cutbacks in economic investments and public infrastructure. This reinforces decline of population and jobs"* (GATZWEILER et al. 2003:564; author's own translation). This 'downward spiral of shrinkage' (see Fig. 1) leads to an increase of vacant tenements and brownfields, and abandonment of public facilities. During the late 1980s several researchers pointed out that the 'shrinking city' could have the potential to become a new type of urban development (HÄUSSERMANN & SIEBEL 1987). They demonstrated their considerations with ongoing processes in cities in West Germany whose economic basis was concentrated on classic industries like ship-building, coal mining and the iron and steel industries (so-called old-industrial cities). Urban politics researchers started an (ongoing) lively discussion about processes of shrinkage and shrinking cities in reaction to the widespread long-term vacancy of tenements in Eastern Germany's cities that occurred in the 1990s as a result of job losses and migration.

During the last fifteen years shrinkage has severely affected small and mid-sized towns in Eastern Germany (the former German Democratic Republic, GDR).[2] They are facing the dramatic economic and social costs of German reunification regardless of their location, specific economic basis, history and administrative status. Today it is widely accepted that the transformation of the Eastern German economy from a planned system to a free market led to a collapse in which the economic structure

2 With the exception of some growing communities in urban peripheries of a few larger cities like Berlin, Dresden, Leipzig and Magdeburg.

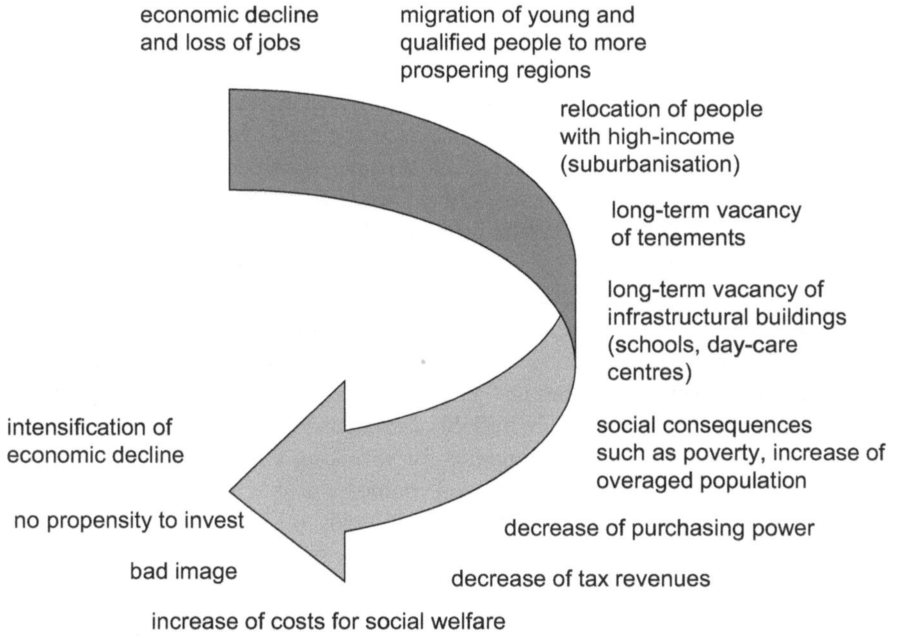

Fig. 1: Downward spiral of urban shrinkage (Beer & Urbane Projekte 2001:25; author's own translation)

completely disintegrated. Industrial decline was accompanied by abandonment of farming cooperatives, liquidation of administrative structures and abandonment of military bases and structures (Hannemann 2004). The result of these simultaneous cutbacks was an immense loss of jobs which could not be absorbed by subsequent economic developments. Economic activity in the *New Länder* (i.e. East Germany) is very low and unemployment has consolidated at an average rate of 20 per cent (status quo in 2006, in comparison to an average unemployment rate of 10 per cent in the West German *Old Länder*). Furthermore, economic decline brought about population shrinkage. Loss of jobs provoked an ongoing wave of labour-related migration to more prosperous regions in Western Germany. Between 1991 and 2004 approximately 1.3 million people relocated predominantly to the western parts of Germany. Population losses of between 20 and 25 per cent per city are not uncommon. Additionly, the birth rate in Eastern Germany fell steeply after 1990: Within six years (between 1988 and 1994) the number of annual newborns dropped by around 60 per cent. And last but not least, further decline of population was caused by soaring processes of suburbanisation. Encouraged by public grants, new housing estates were established in almost every urban periphery in Eastern Germany.

Given these interconnections between different developments, some authors refer to new types of shrinkage processes in Eastern Germany. According to HANNEMANN, shrinkage constitutes *"a downward spiral of structural shrinkage affecting all urban processes"* (HANNEMANN 2004:97; author's own translation). Contrary to transformation in Western Germany's old-industrialised regions, which can be labelled as *structural change*, in Eastern Germany a sudden *structural break* took place (GLOCK 2006). Furthermore, shrinkage in western regions led to processes concentrated in specific locations with region-specific problems. In Eastern Germany, however, it is an area-wide phenomenon.

What concrete consequences does shrinkage have for current urban development? Relating to the built substance of any city, shrinkage firstly indicates less users and a corresponding decrease in demand for uses. This does not automatically mean a deficiency – rather that shrinkage can instigate new themes for urban development such as low rent, more public parks, green fields or spaces for unconventional activities. These aspects may become advantageous for shrinking cities (OSWALT et al. 2002, BBR 2004). However, whether shrinkage becomes a problem for cities depends on different phenomena (GLOCK 2002, POHLAN & WIXFORTH 2005, BERNT 2006):

a) *Scope of shrinkage*: People living in shrinking cities cannot compensate for the decrease in population (and tenants). Affected cities must address the necessity of giving up their vacant buildings. Consequently, whole housing estates were demolished and replaced by forests, lawns or car parks in many East German cities during recent years.
b) *Changed requirements for remaining housing stock*: An increasing proportion of elderly people leads to smaller households and a growing need for furnishing dwellings according to the needs of the elderly. The housing stock no longer suits the requirements of current and prospective residents. As a consequence, the remaining residential buildings in shrinking cities must be rebuilt and enhanced.
c) *New problems within public infrastructure*: Less residents not only means less tenants, but also a decline in use of district heating, a decline in water consumption and less passengers on local public transport. Thus, a massive decrease of population leads to serious problems within different public infrastructure systems and their utilisation. Consequently, in shrinking cities the capacity of infrastructural facilities must be adjusted (see WESTPHAL later in this book).
d) *Expansion of open spaces*: As a result of large-scale demolitions open spaces emerge for which new uses must be found. This is a considerable problem, not an advantage as one might think, for shrinking cities: New inhabitants cannot usually bear the increasing costs for enhancing and operating these new spaces (see RÖSSLER later in this book). Expansion of open spaces and brownfields is accompanied by a decline of potential users and op-

portunities for use. This in turn demands a change in how green spaces and brownfield sites in shrinking cities are dealt with. New types of purpose must be conceived which suit changed conditions due to shrinkage while allowing an aesthetic, functional, and ecologically adequate development.

e) *Decrease of municipal revenues*: The financial scope of shrinking cities is decreasing. Loss of population leads to less municipal revenue. Reduction in revenue on the income side is not compensated for by lower costs on the spending side: In some respects municipal costs remain constant despite a drop in population (e.g. operational costs for a school must be paid regardless of the number of pupils). In other respects costs are even rising within the framework of shrinkage (e.g. for maintenance of technical infrastructure systems).

A political reaction to processes of urban shrinkage occurred in Germany in 2000/2001. The report "Structural changes to housing and the housing industry in the New Länder" (*Wohnungswirtschaftlicher Strukturwandel in den neuen Bundesländern*) was published by a commission of experts (Kommission 2000). Addressing an estimated long-term vacancy of more than one million dwellings in Eastern Germany, mainly in large housing estates, built with pre-cast concrete slabs (*Plattenbauten*) in the 1970s and 1980s, the think-tank recommended the development of a new federal support programme for demolishing vacant tenements. As a consequence, the then German Federal Government launched a new programme called *Stadtumbau Ost* in 2001 ('Urban Restructuring East'; this programme funds large-scale demolition of residential buildings; see Bernt later in this book). In 2004 it was complemented by a second programme named *Stadtumbau West* which focuses on specific processes of shrinkage in some West German regions.

3. Shrinking cities in a global perspective

Although the discussion about shrinkage and shrinking cities is dominated by German researchers, shrinkage is a world-wide phenomenon (Oswalt 2004). Shrinking cities occur worldwide, with regional foci (see Fig. 2). The highest percentage – measured in the terms of population losses – can be found in Europe and North America. Here, losses from 30 to over 50 per cent occurred in some affected cities during recent decades.

When measuring shrinkage in terms of population loss, the identified regions differ according to the spatial scale chosen, as Pallagst & Wiechmann (2005) show for the United States.

In contrast to research in Germany, not only economic and demographic causes for shrinkage are being discussed, but also environmental and political causes. (Oswalt & Rieniets 2006). Often, the causes are intertwined and cannot be sepa-

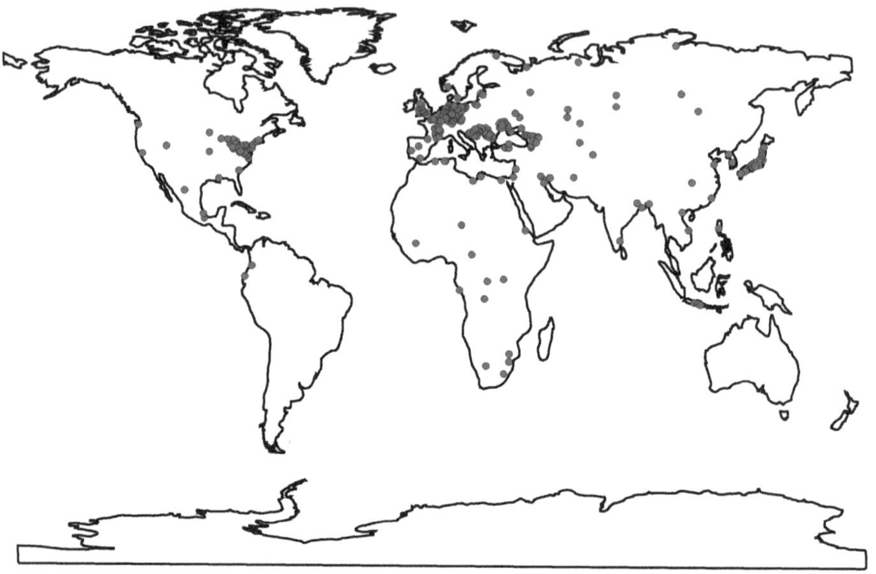

Fig. 2: Shrinking cities worldwide. Plotted cities have shown a loss in population over 10 per cent between 1950 and 2000

rated from one another, and the causes for shrinkage vary from region to region or between cities within a state. These four causes are discussed in more detail below.

a) On a global scale, *economic reasons* for shrinkage and the resulting population migration are predominant over demographic shrinkage of population. There are several economical reasons for shrinkage: The significant reasons for the highest losses are structural changes in economic development, such as the transition from Fordism to Post-Fordism, new technologies and de-industrialisation. Most shrinking cities in Europe and North America are old-industrial cities, now facing the loss of employment and migration of workers towards economically growing areas. Well known examples are the *Ruhrgebiet* in Western Germany and the *Midlands* in Great Britain, both former coal mining regions; the so called *Motor Town* Detroit and the *Midwest Rust Belt* region in the United States. Other economic reasons for shrinkage are exhausted deposits of resources as can be observed in mining cities in Mexico (VARGAS-HERNADÉZ 2007).
b) Corresponding to the sketched German discourse above, in international respects *demographic reasons* for shrinkage are low fertility rates or migration,

or a combination of both. Low fertility rates combined with migration of young people (especially women) and rising life expectancy result in an 'over ageing' of the population and further decline of birth rates: Where potential mothers are absent, no children are born. For example, low fertility rates can be found in Germany and in Japan (OSWALT & RIENIETS 2006:130-131). In contrast, in the US the main reason for shrinkage is migration (PALLAGST & WIECHMANN 2005:106).

c) The third cause discussed in the international debate on shrinkage is *environmental reasons* – e.g. natural hazards, water scarcity or man-made disasters. Natural hazards like earthquakes or hurricanes are generally seen as disasters as soon as they occur in densely populated regions with low protection against these incidents (a current example is the effect of 2005's hurricane *Katrina* on New Orleans).

d) Beside this, *political reasons* like war or political transformation are also an important human-driven reason for shrinkage. Many cities in Eastern Europe's former socialist states are shrinking: Nowadays, they are facing economic transitions and higher mobility options for the inhabitants (RIENIETS 2006).

At a global scale the described reasons for shrinkage usually affect entire regions, not only cities. As a result, most strategies focus on a regional scale. The potential for specific strategies to cope with shrinkage varies, depending on the general political and economical framework as well as on overall societal concepts. Whereas in Germany the Federal Government, the *Bundesländer* and the local municipalities are engaged in strategies coping with urban shrinkage, in the United States implementation tends to be focussed on a regional scale. Developing countries often lack the resources to manage shrinkage processes actively.

4. What is contradictory about shrinking metropolitan regions? The example of the *Berlin-Brandenburg Metropolitan Region*

Growth and shrinkage happen in different spatial contexts; urban regions present the best examples of both occurring together. The *Berlin-Brandenburg Metropolitan Region (BBMR)* has undergone a fundamental political and economic transformation since 1990. The fall of the Berlin Wall and the transformation of Eastern Germany's economy were followed by several principal causes of shrinkage – massive de-industrialisation, demographic change and out-migration of the population to the surrounding countryside. Berlin had approximately 3.4 million inhabitants in 2005 – around 39,000 less than in 1990. The population briefly increased immediately after the fall of the Berlin Wall, peaking at 3.475 million in 1993. Until the late 1990s the population development was dominated firstly by a high general turnover and secondly by suburbanisation processes and the related movement of households into surrounding areas. Between 1993 and 2000 Berlin lost around 93,000 inhabitants, but during the last seven years the population of Berlin has stabilised (Berlin Regional Statistical Office 2007). The population in different areas of Berlin developed disparately between 1992 and 2005. There are areas with a growing population immediately adjacent to those with a shrinking population (Fig. 3 shows the population development on county scale). Most inner city areas of Berlin, and especially districts characterised by large estates of pre-cast concrete slab buildings *(Plattenbauten)*, are decreasing in population, while single-family housing areas are still growing (Senatsverwaltung für Stadtentwicklung 2004). Berlin's district *Marzahn-Hellersdorf,* for example, has to deal with effects of shrinking. *Marzahn* is situated in the former East Berlin and is the largest *Plattenbauten* estate in Germany (about 60,000 dwellings). The district lost approximately 20 per cent of its population from 1992 to 2005 (Bezirksamt Marzahn-Hellersdorf 2007). This had deep impacts on urban and district development: If the decision is made to demolish or deconstruct estates and infrastructure, then an effective means of re-developing the emerging brownfields must be established. The main target of Berlin's efforts in urban restructuring within the above mentioned support programme *Stadtumbau Ost* is *Marzahn-Hellersdorf.*

Simultaneous processes of population growth and shrinkage can also be seen at the municipal level in the suburbs. Nowadays, older suburbs are facing shrinking processes as well (Hesse 2006). Beyond Berlin's city edge, suburbanisation predominantly occurred in the Brandenburg part of the so-called *Sphere of Mutual Influence (SMI) of Berlin-Brandenburg*, accelerating after being hindered by the political situation as well as the physical containment of the Berlin Wall before 1990. From 1992 to 2005, the urban fringe gained around 225,000 inhabitants, a growth of approximately 30 per cent (Brandenburg Regional Statistical Office 2006). The suburban middle-sized town of Falkensee was the fastest growing community nationwide during the 1990s. On the other hand there are communities in the suburbs of Berlin that show population decline (see Fig. 3).[3] Shrinking municipalities of suburban areas experienced processes of people out-migrating from large *Plattenbauten* into single-family houses; nowadays, different from Berlin, these suburbs are affected by decreasing demand for new single houses. Regarding the patchwork pattern shown in Fig. 3 that results from this differentiated distribution of growth and decline, one could agree with Matthiesen (2002) that this is not a uniform, ever-growing suburbanised commuter belt ('*Speckgürtel*') but rather a space characterised by suburban squares ('*Speckwürfel*').

3 Urban planners in shrinking communities do not perceive their communities as 'shrinking' – they either see things as 'stable' or, optimistically, as some sort of intermediate or preparatory stage before resumption of growth.

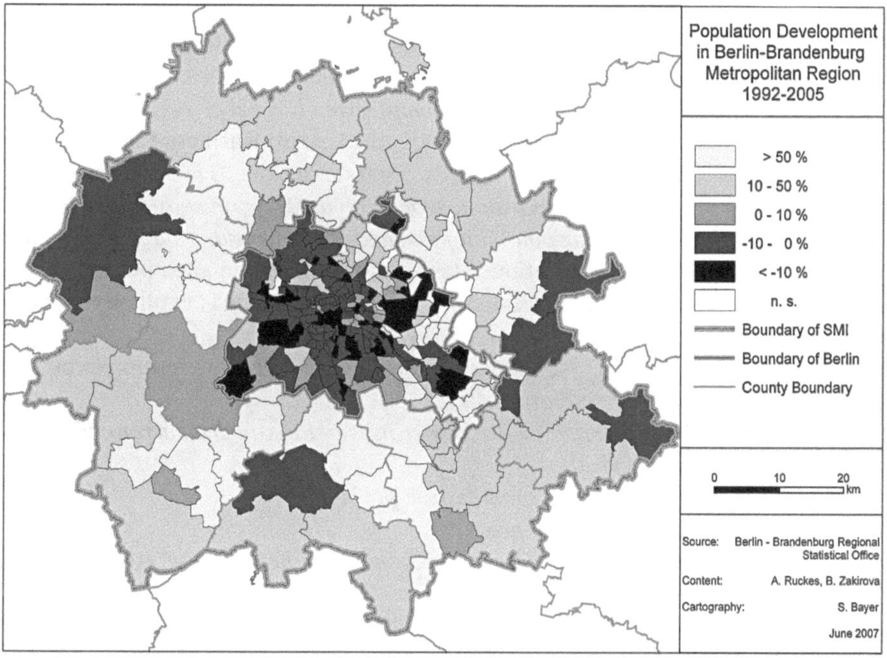

Fig. 3: Population development in the *Berlin-Brandenburg Metropolitan Region* 1992-2005 (own calculations based on data from BERLIN REGIONAL STATISTICAL OFFICE – personal communication 2006 – and from BRANDENBURG REGIONAL STATISTICAL OFFICE – personal communication 2006)

The coexistence of shrinkage and growth in terms of population development within the *Berlin-Brandenburg Metropolitan Region* illustrates the fact that it is a fragmented region. Despite the fact that total population figures in the metropolitan region are stagnating (although today slightly growing), there are obvious areas of shrinkage. The region clearly shows spatial differences in shrinkage phenomena. Thus, it is an excellent laboratory for studying simultaneous processes of urban growth and shrinkage in regard to urban ecology.

5. Brownfields

Processes of shrinkage lead to a change in urban structure; of which the most significant feature is the emergence of vacant lots with different qualities. Areas that are completely abandoned can be found. Alongside these are sites with a decreased intensity of use or sites that will be used in a new way. Within the context of urban ecology, it is useful to call these areas ***brownfields*** or ***brownfield sites***.

The term 'brownfields' was first used in the USA in the early 1990s (THORNTON et al. 2007). The U.S. Environmental Protection Agency (EPA[4]) defined brownfields in a memorandum as "*abandoned, idled, or under-used industrial or commercial facilities where expansion or redevelopment is complicated by real or perceived environmental contamination*" (EPA 1996:1). In 2002, the "Small Business Liability Relief and Brownfields Revitalization Act"[5] gave a legal definition: "*The term 'brownfield site' means real property, the expansion, redevelopment, or reuse of which may be complicated by the presence or potential presence of a hazardous substance, pollutant, or contaminant.*" This definition refers not only to urban sites but to polluted physical properties in general. For example, mine-scarred land is explicitly included in that definition.

In Europe, the CABERNET[6] Network provided a very helpful definition to be used in urban ecology, based on five properties:

- "*sites that have been affected by the former uses of the site and surrounding land;*
- *are derelict and underused;*
- *may have real or perceived contamination problems;*
- *are mainly in developed urban areas;*
- *and require intervention to bring them back to beneficial use*" (OLIVER et al. 2005:1).

Based on this definition, several categories of brownfield sites can be distinguished. For example, for the purpose of brownfield regeneration three groups can be identified: viable sites, marginally non-viable sites and non-viable sites (FRANZ et al. 2006).

The effects of shrinkage on urban ecology on brownfield sites can be investigated using the three factors given in Fig. 4. Various processes occur on brownfields, leading to various positive or negative consequences for city residents. Through city planning there is the opportunity to improve the positive effects of brownfield sites. In the next section, examples from different scientific disciplines will show this in a more explicit way.

6. Potentials of brownfield sites from different disciplines' points of view

Regarding **Economic Geography,** urban brownfields are interesting locations for economic development. Shrinking cities are mostly characterised by a transformation from industry to the service sector. Emerging brownfields can provide a location for the service sector with office buildings or retail industry. Even old buildings on

4 U.S. Environmental Protection Agency

5 Public Law 107-118 (H.R. 2869)

6 Concerted Action on Brownfield and Economic Regeneration Network

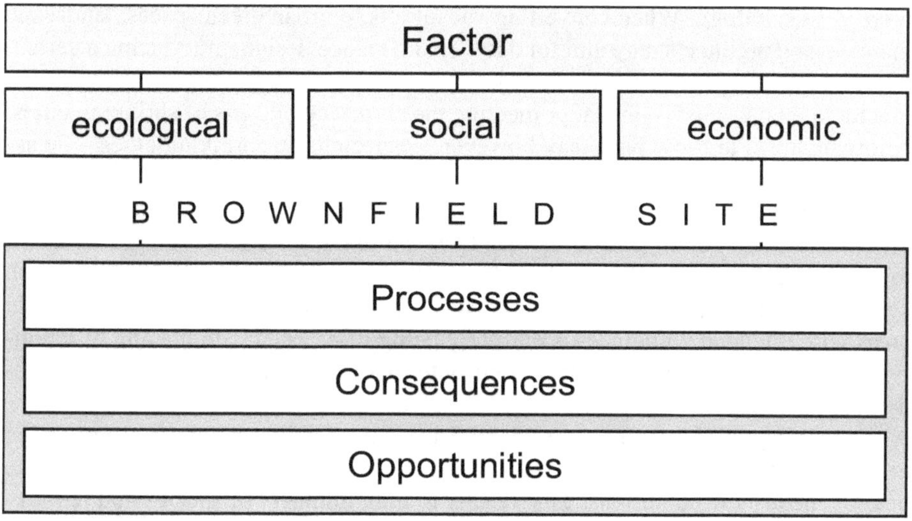

Fig. 4: Dimensions of urban ecology research on brownfield sites in shrinking cities

these vacant sites – e.g. old industrial buildings – are a popular location for services, especially offices, because of their special charm and image. On the other hand, even if the vacant lots are re-used as parkland or similar, they are an important 'soft' location factor for the service industry, especially offices, because green re-use improves the surroundings.

In the context of shrinking cities, the **Urban Geography** field is concerned with how alterations to the functional city structure affect the overall concept of the sustainable European City, with its mixed-use quarters and everyday life of city dwellers. Re-use of brownfield or permanent vacant sites can induce an alteration in the functional city structure – because of addition or omission of city functions. In shrinking cities, whole city quarters as well as single plots, are subject to this process, which intensifies the effect of alteration. When determining the re-use of a large vacant lot in the inner-city, planners try to influence the functional mix of the development by including space for housing units, shopping, leisure facilities and offices. However, the success of a development depends on consumer behaviour as well as economic factors, hence the city dwellers themselves and their action space: If they do not incorporate the newly developed site into their everyday life, the concept will not work.

In shrinking cities, many vacant lots have the potential of becoming green spaces. As such they can enhance residential satisfaction – an important concept in **Environ-**

mental Psychology. When converting vacant lots to urban green spaces, landscape planners and architects may aim for a low maintenance design, introducing a certain degree of wilderness that provides important ecological advantages and promises low maintenance costs – the latter meeting the shrinking budgets of shrinking cities. Allowing areas to grow wild may however – especially in shrinking cities – be associated with recession and downfall.

For potential users to accept urban green spaces some general prerequisites must be met – they must correspond to the needs of the users, they must not pose a threat to perceived security, and they should be aesthetically pleasing (e.g. KAPLAN et al. 1998). How these factors interact with each other and with intrapersonal factors in areas with intended wilderness is currently being researched with the aim of identifying simple design measures that contribute to a greater acceptance of urban wilderness.

A wide variety of **plants and animals** find an array of different habitats on unsealed brownfield sites where various former uses have formed a landscape which provides heterogeneous site conditions. This results in high numbers of species and remarkable occurrences of threatened species (REBELE & DETTMAR 1996).

Natural processes and mechanisms occur in plants, such as colonisation, establishment, growth and decline. Depending on the development time of a site after use, these processes lead to different successional stages dominated by annuals, perennials, shrubs, or trees. Urban brownfields, especially in early successional stages (characterised by short and fragmentary vegetation and open soil patches), provide compensatory living space for open-land species of mammals, birds, reptiles, and insects.

Furthermore, brownfield sites are important as stepping-stone habitats for plants and animals. For example, depending on their spatial and functional coherence, they link the city and its surroundings or different sites inside the city. Brownfields must be considered as very important sites for the protection of the diversity of urban species and habitats as well as for the provision of ecosystem services in cities.

Soil is the integrator between atmosphere, biosphere, lithosphere, and hydrosphere. Whereas natural soil is the result of interaction between these spheres in space and time, **urban soils** are strongly influenced by human activity. The most significant influence is sealing of the soil, which impedes the interaction of the spheres; if soil is sealed it can no longer be a habitat for flora and fauna, nor can it retain water. During rainfall, surface run-off contributes to the pollution of surface water bodies. Sealing the soil on a site also inhibits natural evapotranspiration, which adds to the effect of the urban heat island (WESSOLEK 2001).

Assuming that the area is not contaminated in any way the re-use of brownfields resulting from the shrinkage of cities only affects soil if it is unsealed.

The unsealing of soil may go hand-in-hand with the creation of urban green space, as a permanent solution. Positive side effects include the decrease of surface run-off and increase of evapotranspiration, thus an improved local climate as well as the decongestion of built structures. Interim solutions or new constructions should consider the role of soil in the urban water balance and aspire to concepts that include pervious sealings allowing for infiltration of rainfall. **Urban watercourses** are influenced by various factors, such as pollution from domestic and industrial sewage waters, abrasion from streets and vehicles, and polyaromatic hydrocarbons from boats and their protective coatings (Papiri et al. 2003, Mahler et al. 2005). Natural habitats in river banks are diminished by sheet pile walls and siltation (Paul & Meyer 2001). Hence, preservation of sustainable populations of limnic organisms is impeded. These reasons account for the characteristic lack of sensitive native species in many urban waters (Walsh 2006).
Regarding watercourses in shrinking cities, it is to point out that besides the typical urban set of problems further impact by bordering brownfields with former industrial use and therefore mostly contamination of soils might be expected. Eluviations of contaminants into surface waters can occur and cause additional stress for limnic organisms, or even lead to dislocation of pollution into groundwater layers. Bordering watercourses are often disregarded when re-naturation programs are implemented at brownfield sites. For further development of research on possible remote damage by brownfields the influence on bordering surface waters, with emphasis on the limnic organisms and the coherences with groundwater layers, should be examined. Due to human activity urbanised areas are hotspots of **air pollution** from various sources, e.g. traffic, heating and industry. In industrialised countries pollutants emitted by traffic – like nitrogen oxides, organic compounds and fine particles – are of special interest in urban agglomerations (Fenger 1999).
Brownfield sites may emit harmful gasses or particles depending on their former use. Conversely, reduced human activity causes a decrease of anthropogenic emissions typical of urban environments. Emissions from nearby sources are dispersed over the area leading to a dilution and therefore lower concentration of pollution. Vegetation-covered brownfields may improve air quality since leaf surfaces provide an increased deposit area for gaseous or particulate pollutants. This effect is subject of ongoing research.
Urban form and air quality are also linked (Marquez & Smith 1999). City planners must consider that dispersed locations of industrial activities, residential areas, business districts and brownfield sites may induce an increase in transport of goods and traffic, which affect urban air quality. Therefore, from the point of view of urban air quality, the pattern of brownfield sites in shrinking cities should not originate randomly but be part of an integrated planning procedure.

The pattern of brownfields within a city and the spatio-temporal development of land cover types on single urban brownfields have impacts not only on air quality but also on other parts of the urban ecosystem e.g., urban climate and urban hydrology and therefore consequences on the well-being of urban residents. Hence, approaches to include brownfield management in an overall sustainable urban development require appropriate information of the distribution and actual land cover types like sealed soils or different stages of vegetation succession on brownfields. As brownfields cannot be detected solely by **remote sensing** techniques, utilizing remote sensing imagery in combination with auxiliary GIS data allows for recognition of brownfield sites. Remote sensing subsequently provides a tool for analysis of the actual condition of these sites and retrospective analysis of land cover change on brownfield sites and their surrounding areas. Therefore, analysis of single plots are possible as well as large areas, leading to a more synoptic view of the environmental impact of urban brownfields.

7. Conclusions

Although increasing urbanization is a dominant phenomenon on a global scale, shrinking cities can be found in regions all over the world. Both processes make great demands on scientists and practitioners. As opposed to fast growing urban agglomerations suffering a high pressure on land, shrinkage increases the availability of open space e.g. brownfields that can be subject to urban planning. However, the presence of open space in shrinking cities does not imply the absence of conflicts between different interests, such as nature conservation, recreation and infrastructure. Shrinking cities are characterised by a pattern of brownfield sites on which highly dynamic processes occur, driven by nature and/or human activity. Hence, shrinking cities challenge urban ecologists to adopt new interdisciplinary approaches for understanding the functional interactions between ecological, social and economical processes.

Sustainable development in shrinking cities needs proper evaluation of the different functions that brownfields fulfil in urban environments. Appropriate management of such open spaces may increase quality of life, and will therefore greatly benefit from interdisciplinary research in urban ecology.

References

BBR (Bundesamt für Bauwesen und Raumordnung) (ed.) (2004): Zwischennutzung und neue Freiflächen. Städtische Lebensräume der Zukunft. Berlin.

BEER, I. & URBANE PROJEKTE SCHMITZ POTSDAM (2001): Obere Talsandterrasse in Schwedt/Oder. Integriertes Entwicklungs- und Handlungskonzept im Rahmen des Förderprogramms "Soziale Stadt". Gutachten im Auftrag der Stadt Schwedt/Oder.

BENKE, C. (2005), Historische Schrumpfungsprozesse: Urbane Krise und städtische Selbstbehauptung in der Geschichte. In: GESTRING, N. et al. (eds.): Jahrbuch StadtRegion 2004/2005. Schwerpunkt: Schrumpfende Städte. Wiesbaden: 49-70.

BERLIN REGIONAL STATISTICAL OFFICE (STATISTISCHES LANDESAMT BERLIN) (2007): Statistics. online available at http://www.statistik-berlin.de/framesets/berl.htm (10 July 2007).

BERLINER DEBATTE INITIAL (2002): "Schrumpfende Städte" 13(2).

BERNT, M. (2006): Stadterneuerung, Stadtumbau und Schrumpfung: Was ist neu? In: ALTROCK, UWE et al. (eds.), Sparsamer Staat – Schwache Stadt? Planungsrundschau 13. Berlin: 115-126.

BEZIRKSAMT MARZAHN-HELLERSDORF (ed.) (2007): Basisbericht 2006. Teil 1: Demographie Marzahn-Hellersdorf. Beiträge zur integrierten Gesundheits- und Sozialberichterstattung. Berlin.

BRANDENBURG REGIONAL STATISTICAL OFFICE (LANDESBETRIEB FÜR DATENVERARBEITUNG UND STATISTIK POTSDAM) (2006): Statistische Berichte (1992-2006). Potsdam.

BRANDSTETTER, B., LANG, T. & A. PFEIFER (2005): Umgang mit der schrumpfenden Stadt – ein Debattenüberblick. Berliner Debatte Initial 16(6): 55-68.

EPA (1996): Support of Regional Efforts to Negotiate Prospective Purchaser Agreements (PPAs) at Superfund Sites and Clarification of PPA Guidance. Memorandum, online available at http://www.epa.gov/superfund/programs /reforms/docs/ppa_clar.pdf (05 June 2007).

FENGER, J. (1999): Urban air quality. Atmospheric Environment 33: 4877-4900.

FRANZ, M., PAHLEN, G., NATHANAIL, P., OKUNIEK, N. & A. KOJ (2006): Sustainable development and brownfield regeneration. What defines the quality of derelict land recycling? Environmental Sciences 3(2): 135-151.

GATZWEILER, H.-P., MEYER, K. & A. MILBERT (2003): Schrumpfende Städte in Deutschland? Fakten und Trends. In: BBR (BUNDESAMT FÜR BAUWESEN UND RAUMORDNUNG) (ed.): Informationen zur Raumentwicklung 10/11: 557-574.

GLOCK, B. (2002): Schrumpfende Städte. Berliner Debatte Initial 13(2): 3-10.

GLOCK, B. (2006): Stadtpolitik in schrumpfenden Städten. Duisburg und Leipzig im Vergleich. Wiesbaden.

HANNEMANN, C. (2004): Marginalisierte Städte. Probleme, Differenzierungen und Chancen ostdeutscher Kleinstädte im Schrumpfungsprozess. Berlin.

HÄUSSERMANN, H. & W. SIEBEL (1987): Neue Urbanität. Frankfurt am Main, New York.

HEINEBERG, H. (2005): Cities in Germany – Between Growth, Shrinkage and Restructuring. Geographische Rundschau International Edition 1(1): 30-41.

HESSE, M. (2006): Suburbanisation. In: OSWALT, P. & T. RIENIETS (eds.): Atlas of Shrinking Cities. Ostfildern: 63.

KABISCH, S., BERNT, M. & A. PETER (2004): Stadtumbau unter Schrumpfungsbedingungen. Eine sozialwissenschaftliche Fallstudie. Wiesbaden.

KAPLAN, S., KAPLAN, R. & R.L. RYAN (1998). With People in Mind – Design and Management of Everyday Nature. Washington, D.C.

KAUFMANN, F.-X. (2005): Schrumpfende Gesellschaft. Vom Bevölkerungsrückgang und seinen Folgen. Frankfurt am Main.

KOMMISSION "WOHNUNGSWIRTSCHAFTLICHER STRUKTURWANDEL" (2000): Wohnungswirtschaftlicher Strukturwandel in den neuen Bundesländern. Bericht der Kommission im Auftrag des Bundesministeriums für Verkehr, Bau- und Wohnungswesen. Berlin.

LANG, T. & E. TENZ (2003): Von der schrumpfenden Stadt zur Lean City. Prozesse und Auswirkungen der Stadtschrumpfung in Ostdeutschland und deren Bewältigung. Dortmund.

MAHLER, B.J., VAN METRE, P.C., BASHARA, T.J., WILSON, J.T. & D.A. JOHNS (2005): Parking lot sealcoat: An unrecognized source of urban polycyclic aromatic hydrocarbons. Environmental Science & Technology 39: 5560-5566.

MARQUEZ, L.O. & N.C. SMITH (1999): A framework for linking urban form and air quality. Environmental Modelling & Software 14: 541–548.

MATTHIESEN, U. (ed.) (2002): An den Rändern der deutschen Hauptstadt. Suburbanisierungsprozesse, Milieubildungen und biographische Muster in der Metropolregion Berlin-Brandenburg. Opladen.

OLIVER, L., FERBER, U., GRIMSKI, D. & P. NATHANAIL (2005): The Scale and Nature of European Brownfields. CABERNET 2005 – International Conference on Managing Urban, 13-15 April, Belfast, Northern Ireland, UK. Nottingham. Online available at http://www.cabernet.org.uk/ resourcefs/417.pdf (05 June 2007).

OSWALT, P. (ed.) (2004): Schrumpfende Städte. Band 1: Internationale Untersuchung. Ostfildern-Ruit.

OSWALT, P., OVERMEYER, K. & W. PRIGGE (2002): Experiment und Utopie im Stadtumbau Ostdeutschlands. Berliner Debatte Initial 13(2): 57-63.

OSWALT, P. & T. RIENIETS (2006): Atlas of Shrinking Cities. Ostfildern.

PALLAGST, K. & T. WIECHMANN (2005): Shrinking Smart? Städtische Schrumpfungsprozesse in den USA. In: GESTRING, N. et al. (eds.), Jahrbuch StadtRegion 2004/05: Schwerpunkt: Schrumpfende Städte. Wiesbaden: 105-127.

PAPIRI, S., CIAPONI, C., CAPODAGLIO, A., COLLIVIGNARELLI, C., BERTANZA, G., SWARTLING, F., CROW, M., FANTOZZI, M. & P. VALCHER (2003): Field monitoring and evaluation of innovative solutions for cleaning storm water runoff. Water Science and Technology 47: 327-334.

PAUL, M.J. & J.L. MEYER (2001): Streams in the urban landscape. Annual Review of Ecology and Systematics 32: 333-365.

POHLAN, J. & J. WIXFORTH (2005): Schrumpfung, Stagnation, Wachstum – Auswirkungen auf städtische Finanzlagen in Deutschland. In: GESTRING, N. et al. (eds.), Jahrbuch StadtRegion 2004/05: Schwerpunkt: Schrumpfende Städte. Wiesbaden: 19-48.

REBELE F. & J. DETTMAR (1996) Industriebrachen. Stuttgart.

RIENIETS, T. (2006): Urban Shrinkage. In: OSWALT, P. & T. RIENIETS (eds.): Atlas of Shrinking Cities. Ostfildern: 30.

SENATSVERWALTUNG FÜR STADTENTWICKLUNG BERLIN (ed.) (2004): Bevölkerungsprognose für Berlin 2002-2020. Bevölkerungsentwicklung in der Metropolregion Berlin 2002-2020. Berlin.

THORNTON, G., FRANZ, M., EDWARDS, D., PAHLEN, G. & P. NATHANAIL (2007): The challenge of sustainability: incentives for brownfield regeneration in Europe. Environmental Science & Policy 10(2): 116-134.

VARGAS-HERNANDÉZ, J.G. (2007): Verschüttete Entwicklung. Das Schrumpfen von Bergbaustädten in Mexiko am Beispiel von Cerro de San Pedro. Berliner Debatte Initial 18(1): 36-44.

WALSH, C.J. (2006): Biological indicators of stream health using macroinvertebrate assemblage composition: a comparison of sensitivity to an urban gradient. Marine and Freshwater Research 57: 37-47.

WESSOLEK, G. (2001): Bodenüberformung und -versiegelung. Handbuch der Bodenkunde, 11. Erg. Lfg. 04/01: 1-29.

URBAN BROWNFIELDS AS TEMPORARY HABITATS: STRATEGIES OF PLANTS FOR SURVIVAL IN A CHANGING ENVIRONMENT

Ingeborg Schinninger, Rudolf Maier

Green and colourful from spring until early autumn, brown the rest of the year, this is the fate of urban brownfields[1]. Their long-standing development is often stopped by local economic interests. However, the longevity of urban brownfields which bridge the period between demolition and erection of buildings, becomes even more critical as the process of shrinkage affects compact town centres.

Brownfields represent ideal urban green spaces for cross-disciplinary study among a variety of professions, including economists and urban planners who focus on space utilisation, and ecologists who study biocoenoses as well as the broader implications of sustainability. Citizens are also important stakeholders in brownfield sites, recognizing that spontaneous vegetation balances and enhances the environment; the appearance of brownfields is not static, but fluctuates with the seasons and varies as land uses change.

There are many expectations brownfield plants must meet, such as representing urban green-spaces, serving as urban corridors, and acting as nuclei of biodiversity. Additionally, they should awaken an understanding and reverence for nature within the local populations while creating an aesthetically pleasing environment. In addition to these demands, plants are exposed to environmental stress factors for a large part of the year.

Urban climate differs from the climate of the surrounding countryside in some respects (i.e., Kratzer 1956, Sukopp et al. 1974, Gilbert 1994, Kuttler 1998). In general towns are heat islands in the landscape; the town itself is again characterized by smaller thermal islands that develop from massive human influence in the centers. In shrinking cities, however, the degree of human influence is changing. A positive correlation between number of inhabitants and heat islands (Kuttler 1998) creates an expectation of local changes of urban temperatures in shrinking cities. Connected to this is the potential change in urban flora due to the number of vascular plant species in Central European towns being more closely correlated with the number and density of inhabitants than with the urban area (Brandes & Zacharias 1990).
Not only can a change in the number of city dwellers affect the urban climate, but also it can be affected by changes in global temperature caused by climate change.

1 For definition of the *term* brownfield see chapter 2.

Another important aspect is the question of whether and how global warming will affect cold hardening or heat resistance of plants growing in urban environments which are already warmer than the surrounding countryside.

Besides well-maintained green spaces and parks, plants of urban ecosystems mainly survive in areas of temporary abandonment formerly used for industry, business and/or housing. These habitats arise from the continual change of urban infrastructure and also as a consequence of the phenomenon of shrinking cities.
These areas of various sizes are re-established after a certain period; nevertheless there is enough time for plants to populate them. Some species are more 'efficient' at settling in quickly to newly disused areas. Other species are able to adapt well to prevailing disturbance (mechanical disturbance i.e., construction work and use as playgrounds or increased nitrogen addition due to dogs). Additionally, other species possess distinct mechanisms to 'defend' their demands for habitat at an advanced successional stage.

Spontaneous development of vegetation occurs on urban brownfields despite conflicting ecological factors. Vegetation is determined by the quantity of diaspores in the soil and from the addition of diaspores from the surroundings. The increased number of transportation corridors (highways, railways, ships and aeroplanes) and the ease of movement between borders and continents brought about by globalisation, has strengthened the migration of plants from areas with similar climates; towns have become rich in neophytes, among which are many plants originating from areas with warm or dry-hot summers. The urban climate is very convenient for these species. For example, *Chenopodium botrys* is common throughout Southeast Europe to East Asia, *Senecio inaequidens* comes from South Africa, *Datura stramonium* from Mexico and *Ailanthus altissima* is native to China (Essl & Rabitsch 2002). These alien species sometimes spread aggressively and become invasive. On the one hand their number is enhanced by urbanisation (Kowarik 1995), on the other hand these species have effective strategies to make use of temporary urban habitats and 'conquer' new areas. Natural enemies are often absent, the alien species occupy 'empty niches', are often aggressive, and are well-adapted to disturbances (Callaway & Hiero 2006).

Adaptation to the warmer and drier urban climate can also be seen in indigenous urban plants. Above all, urban brownfields distinguish themselves by their supply of resources as reservoirs for these plants, so that systems with high biodiversity can develop.

1. Sucession and adaption strategies to biotic factors

An urban ecosystem provides habitats for many plant species that originate from different habitats, colonising town and spreading through a variety of adaptation strategies. Plant species that benefit from significant genetic variation and phenotypic plasticity (individually morphological and physiological modifications) are able to adapt well to environmental influences (Bradshaw 1984, Bazzaz 1986, Joshi et al. 2001).

In ecological research there are many theories and models to explain and describe adaptation strategies through various life cycle phases of plants as well as investigate their solutions to environmental changes and therefore to subdivide species in functional groups. Well-known examples of such theories are the r – K life history theory of MacArthur (1962) and MacArthur & Wilson (1967), or the three strategy model – competitive strategy, stress-tolerant strategy, and ruderal strategy (CRS-model) – proposed by Grime (1977, 2001).

Plants possess a number of adaptation strategies for surviving often heterogeneous site conditions and their differing stress situations due to abiotic and biotic environmental factors. From an evolutionary standpoint, a plant is prepared with a certain ecological potency against ecological site conditions. However, optima and pessima are adjusted to local conditions within the scope of their modification possibilities (in other words, acclimatisation). Plants respond to rapidly varying site conditions, i.e., light-and-shadow-play within a plant stand with rapid and reversible modulatory adaptations such as stomatal opening and closure (Larcher 2003).
Every stress produces a strain – a reaction to stress – to which the plant reacts with avoidance and tolerance strategies (Levitt 1980). When the avoidance strategies are exhausted, the plasmatic tolerance – the resistance of the plasma – is decisive and responsible for the survival on the site. Tolerance strategies are energy consuming and are only used in urgency (i.e., phytochrome- and temperature-induced acquisition of wintry frost resistance in Northern latitudes).

Urban brownfields are generally highly radiated, hot and dry locations. Protective mechanisms against photoinhibition, photodestruction, heat and drought resistance must be developed. The sites are colder than the surrounding countryside during winter, caused by higher irradiation from the relatively bare ground. Therefore strategies against cold are necessary.

In fact, the ecological conditions within a plant population are subject to changes (Bazzaz 1996), like reduction and spectral modification of solar radiation, reduction of temperature by shade effects, change of transpiration and diffusion resistance of leaves caused by wind-reducing effects of plant stands. Nevertheless, temporal green areas are classified as sun-types. In urban empty spaces and brownfields a species-rich vegetation mosaic appears, ranging from pioneer stages to tree stands, according to age and geographical position of the abandoned area. If a space is used for recreation, especially as a playground for children, open areas are occasionally formed on which vegetation development can recommence, as low-competition pioneer plants can survive.

The progress of succession in urban areas is well documented (i.e., by Bornkamm 1986, Kowarik 1986, Pysek & Pysek 1991), and also in disused urban railway areas (i.e., by Sukopp et al. 1974, Kowarik & Langer 1994, Prasse & Ristow 1995). The prediction of successional dynamics and the sequence of biocoenoses in urban ecosystems is, because of multiple human disturbances, more difficult than for non-urban ecosystems (Trepl 1994).

Plant populations are in a dynamic, rather than static state. Poorly competitive pioneer or initial stages of plant colonisation are replaced by successive colonists in the course of the time, until a relatively stable community, or climax, is reached. Meanwhile, the successional development is occasionally reduced to the initial stages by disturbances.
Initial stages are normally characterised by low biodiversity with a high number of individual plants and a low biomass production. The succeeding stages exhibit a higher diversity and an increased biomass production.

Unlike species occurring in later successional stages, species of pioneer stands have developed different ecological strategies to successfully colonise a new site.

Growth and dispersal strategies
The so-called **r-strategists**, dominating the early successional stages, use the rich supply of resources in the soil and have a rapid ontogenesis with a high reproduction potential. By producing many seeds they procreate in large numbers. For example, one plant of *Galinsoga* can produce more than 300,000 achenes, whereas the average production is 6,000 achenes (Hegi 1979). In the succeeding stages when resources become scarcer, more and more **K-strategists** appear, whose seed output is quantitatively lower but whose chances of survival are relatively higher. In contrast to r-strategists, K-strategists are longer-lived and have a slower ontogenesis. While pioneer stages exhibit a greater tolerance towards extreme site conditions, variable environmental conditions and disturbances, species of succeeding stages and climax stages are more demanding, occupy more distinct niches, use resources under

competition more effectively and are thus able to out-compete primary colonisers (MacArthur 1962, MacArthur & Wilson 1967, Pianka 1970, Odum 1969). Recent investigations carried out by Godefroid & Koedam (2007) prove that ruderal species that quickly complete their life cycle and produce a high number of seeds are strongly favoured in densely built-up areas.

To colonise an area, a species not only must tolerate prevailing site conditions and possible competition, it also needs an efficient strategy adapted to the particular environmental conditions for the **dispersal of seeds**. While many species produce plenty of seeds of lightweight construction with low seed weight and good dispersal mechanisms using the wind (**anemochory**) (i.e., *Senecio inaequidens*), there are others that invest in mechanisms such as the production of elaiosomes, nutrient rich appendages of diaspores, which are consumed by ants (myrmecochory) (i.e., *Chelidonium maius*, *Corydalis cava*, *Anchusa officinalis*, species of the genus *Centaurea*). Further adaptations to dispersal via animals (**zoochory**), which can effectively cover a long distance and therefore contribute to the colonisation of temporary habitats, are achieved through specific adherence mechanisms (i.e., *Arctium lappa*) or dispersal via digestion (i.e., *Hedera helix*, *Viscum album*).

Once a habitat is colonised, various strategies take effect. Some species reproduce through self-dispersal (**autochorie**): via soil creepers (i.e., *Erodium cicutarium*) or trailing stems (stolons) long-distance dispersal is limited, as is the probability of reaching a new site unsuitable for colonization.

The soil diaspore bank is of great importance for the regenerative capability of areas where human disturbances often occur. Seeds are 'resting stages', they can outlast long periods of time without losing their germination ability. To survive in disturbed habitats, dormancy of seeds may be a better strategy than dispersal (Schippers et al. 2001).

Facilitation

Successional dynamics and community structure are influenced by abiotic and biotic factors, which again undergo changes during colonisation caused by plant cover. **Plant-plant interactions** are substantial biotic factors that exert great influence on species diversity, community structure and ecosystem productivity, and can be positive as well as negative. Positive interferences or facilitation have recently received much interest, (i.e., Callaway 1995, Callaway & Walker 1997, Holmgren et al. 1997, Cardinale et al. 2002). An example of the positive effects of biotic interactions between species is the way that pioneering mosses accumulate humus and store water, enabling colonisation of temporary habitats by vascular plants.

Symbiosis
Another positive bi-system-interaction, which can be viewed as an adaptation strategy of brownfield plants to environmental conditions, is symbiosis with nitrogen fixing organisms (mycorrhiza). Although urban soils may be quite eutrophied and thus contain plants with elevated nitrogen demand (i.e., *Chenopodium album*, *Artemisia vulgaris* or *Sambucus nigra*), species such as *Robinia pseudacacia* (native to North America) benefit from the ability to fix nitrogen on sites with low soil nitrogen content.

Competition
Negative biotic interferences such as competition or competitive exclusion and their effects on resource use, productivity and community structure is an important field of ecological research (i.e., Whittaker 1965, Grime 1973, Tilman 1982, 1988, 1994, Schoener 1983, Goldberg & Barton 1992, Casper & Jackson 1997). As time progresses and plant cover develops, intra- and interspecific competition gain importance. To acquire limited resources, plants develop different strategies. One of these strategies is **allelopathy**, the ability of plants to secrete organic compounds that delay or halt growth or germination of other plants. Allelopathy may become significant as an aggressive competitive strategy within urban plants so that monocultures may develop. Also the Tree-of-Heaven, *Ailanthus altissima*, native to China and proliferating in urban areas, uses phytotoxic effective components (above all ailanthone; Heisey 1990, De Feo 2003), which guarantee invasive success (Bais et al. 2003). In addition, this species has a high production of diaspores for wind dispersal and possesses high regenerative potential via vegetative reproduction (such as root suckers). Besides, *Ailanthus* is sun-adapted, has a high tolerance towards drought and heat (maximum temperatures of 47.8 °C are endured in China), is tolerant towards salt and air pollution and is well adapted to the prevailing water availability of its location. Therefore, this species is well adapted to disturbances and capable of fast colonisation of anthropogenic areas (Kowarik & Böcker 1984, Essl & Rabitsch 2002, Schinninger et al. 2002, Schinninger 2005, Punz et al. 2004; see also review Kowarik & Säumel 2007). Allelopathic substances with a different mode of action (Schlee 1992) exist within pioneer species of secondary successions such as *Ambrosia artemisiifolia*, *Erigeron annuus* and species of the genus *Solidago* (Bohlmann et al. 1973).

Ethereal oils, glycosides (such as juglone in walnut trees), growth regulators, terpenoids (*Salvia verticillata*, *Senecio jacobea*) or phenolic acids have an allelopathic effect. Plants not only gain a competitive advantage in the acquisition of (temporarily) available resources through growth inhibition of other plants, but those constituents also play an important role as defence against mammalian herbivores, specialist insects and fungi (i.e., RICE 1977, SCHLEE 1992, LANGENHEIM 1994, REIGOSA et al. 2006).

Not only do biotic interactions determine the life on urban brownfields, abiotic factors are also of essential importance in shaping these habitats. If in the following the attempt is made to structure adaptations towards abiotic stressors on the basis of ecological factors, it should be noted that many morphological, structural and physiological adaptation strategies apply not only to one single stressor, but to multiple stresses. Only on the physiologic-biochemical level do adaptations become more orientated towards a single stressor. (See also HUANG 2006, MOONEY et al. 1991).

2. Adaption strategies to the urban climate

Solar Radiation

Too much light can cause photoinhibition or even photodamage. A morphological and structural adaptation of xerophytes, which include many plants found in temporary green urban areas, is a reduction of strong light by means of **hairiness** (i.e., *Verbascum sp.*). Bright, dead hairs increase reflection of solar radiation; moreover they have a shading effect and increase boundary layer resistance of the leaf. In this way, leaves reduce the intensity of radiation, remain cooler and transpire less. Also, shiny leaf surfaces through **cuticular waxes** increase reflection to ease the effects of microclimatic extremes

Structural adjustments such as the **wall thickening of epidermal cells** also reduce radiation. The **palisade parenchyma** of sun leaves is mostly multilayered (i.e., *Galium verum*) (KAUSSMANN & SCHIEWER 1989) and acts as a light filter enabling deeper chloroplasts to photosynthesise, while chloroplasts in the uppermost layer of palisade cells may be affected by photoinhibition (LARCHER 2003). Chloroplasts of sun leaves possess thin thylakoid stacks which thicken towards the underside of the leaf (SCHULZE et al. 2002). Finally, chloroplasts also orient themselves according to the light which is classic behaviour of many aquatic plants. Chloroplasts attempt to avoid strong light by means of either a parastrophic position, a light oriented diastrophic position, or a randomly distributed peristrophic position (TROLL & HÖHN 1973).

The reduction of radiation absorbing surface area may also be achieved by movement-based physiological adaptations, through **up-** or **downwards directed leaves** and **leaf rotations** (*Lactuca serriola*), as well as through the **rolling** (i.e., *Hordeum* *murinum*) and **folding of leaves** (i.e., *Poa pratensis*) which can be found in Poaceae (Eschrich 1995) among others.

A number of physiological-biochemical adaptations to light climate exist on the cellular level. Plants growing on sunny sites reduce their **chlorophyll content**. **Anthocyanins** and **flavonoids** serve as a light filter. Plants react to inhibition of photosynthesis and damage of membranes by intense light using fast physiological-biochemical adaptations. Critical moments in a plant's life are those in which the plant is forced to close its stomata by water stress. The light reactions continue, but the regular course through the Calvin cycle is disturbed. Electrons of the chlorophyll that are excited by light accumulate as well as oxygen, causing oxidative damage of the photosynthesis apparatus. By the switch of the CO_2-binding enzyme Rubisco to O_2-fixing, the Calvin cycle can continue in a limited way, while **glycolate metabolism** with **photorespiration** is activated, which releases carbon dioxide (CO_2). Through the consumption of oxygen, energy- and reduction equivalents, plants can counteract the formation of reactive oxygen species (ROS) and radicals. Existing aggressive oxygen species and radicals are detoxified, to this end, **superoxide dismutase** and **ascorbate peroxidase** are important, as are **glutathione** and **α-Tocopherol**. The energy of excited electrons which cannot follow the Z-scheme of photosynthesis is dissipated in the photosystems via fluorescence and heat. Under high light stress the content of carotenes, which have the task of abducting surplus electron energy in the form of heat, is elevated. The built-up surplus of reductive capacity in the form of NADPH (nicotinamide adenine dinucleotide phosphate) is dissipated by the **xanthophyll cycle** (comprehensive description i.e., Lawlor 1990, Richter 1996, Larcher 2003, Schulze et al. 2002).

Finally, the **C4 syndrome** seen in certain Poaceae, Chenopodiaceae and Amaranthaceae among others, that need plenty of light and simultaneous water economy, is an example of an adaptation strategy dealing with the extreme conditions of brownfield sites.

Temperature

Plants that survive the hot summers on radiation-rich sites use adaptations such as **hairiness** and **evasive movements**, which reduce radiation to concurrently prevent heat damages. Further measures to counteract overheating of plant organs caused by absorption of solar radiation during summer are **small**, often **rounded leaves** and sometimes radiation-scattering **round shoots** (*Chondrilla juncea*) which are also able to perform photosynthesis. **Incised leaves** (*Artemisia sp.*) are a common means through which convective processes cooling the leaf increasingly take effect (i.e., Schulze et al. 2002). Also **fluttering of the leaves** enables heat loss, although, for

Fig. 1: Dead leaves of Echium vulgare forming a heat shield to protect younger leaves (22.3.2004)

example, the obvious fluttering of *Populus*-leaves is considered to have a minimal influence on heat transfer (GRACE 1996).

Plants also withstand summer heat by growing leaf buds in the **shade** of the individual plant body. During times of strong solar radiation and dryness some rosette plants curve the lowest leaves around the heat sensitive young leaves (LARCHER 2003). The basal leaves of the rosette of *Echium vulgare* cannot tolerate the heat stress caused by warmed soil; they die and assume the function of a 'heat shield' for the leaves above, thus enabling them to survive (Fig. 1). The heat resistance of living leaves directly above the dead leaf rosette of plants in an abandoned railway area was found to be 40° Celsius, while the soil surface temperature reached 50-53° Celsius (SCHINNINGER 2005).

Transpiration, the release of water vapour via the stomata, has substantial importance for the reduction of leaf temperatures higher than air temperatures. The transpiration of plants growing on urban brownfields may, however, be rather restricted and only be well realized within deep rooting plants.

Summer hardening processes at cellular level increase heat resistance. The biomembranes of cells are gradually adjusted to the prevailing temperature through exchange of fatty acids, a similar process to the acquisition of frost resistance. Sudden intense heat stress leads to a fast adaptation by synthesis of heat-shock proteins at the expense of the expression of housekeeping genes (comprehensive description i.e., LARCHER 2003, SCHULZE et al. 2002).

The acquisition of winter **frost resistance** at cellular level is a gradual process linked to reduction of daylight, growth stop and hardening cold.
Stabilisation of membranes, splitting up of the central vacuole into many small vacuoles, accumulation of substances that delay freezing and associated dehydration, lead to a frost resistant cell. Above all, woody plants act like this, while herbaceous plants, being low-growing, seek the protection of ground level and snow cover.

To start building up frost resistance in autumn, low to near freezing temperatures are required. Towns are heat islands compared to the surrounding countryside. If global warming causes an additional rise in temperature of urban climate, the question arises of how plants on urban brownfields (and in the urban area in general) will be affected in their ability to achieve frost resistance and cold hardening.
Climate change and an increase in urban temperature may not only affect plants in their winter cold hardening, they may also cause additional heat stress and heat damages in summer.

It should be noted that the isolating effects of **cuticle**, **periderm**, **bark** and **aerenchyma** also intervene in the heat balance of plants, as well as dense **leaf sheaths covering the basal buds**, which gain importance by reducing heat loss in winter. **Temporal** and **spatial avoidance**, seen in the **life forms** of Raunkiær (1910, 1934), are strategies that cope with harsh environmental conditions. On abandoned railway areas in Vienna annual species such as *Erophila verna agg*. or *Saxifraga tridactylites*, dominate free space in spring, the time of favourable living conditions, and cover the area with a green layer (Schinninger 2003, 2005). By summer many seeds, which do not germinate before another favourable time arises, have been produced. In active railway areas these species have the advantage of having finished their life cycle before herbicides are spread. Similar strategies apply to geophytes which transfer a large part of their vegetative phase to the soil. However, in abandoned, undisturbed areas annual species are – in the sense of succession – replaced by perennial species, primarily hemicryptophytes with perennating buds just above the soil surface which find shelter beneath litter, enclosed in the remnants of dead leaves or under snow cover. Finally, on urban browfields, there are also phanerophytes, shrubs and trees that shed their foliage in seasonal climate to spend the winter with frost-hardy perennating buds above the snow.

Again, life forms do not simply adapt to a single stressor, but adjust to multiple factors which can induce a stress situation in the plant. This also applies to those plants that avoid intense light, heat and dryness in the shade of trees, shrubs or buildings.

Water

The adjustable stomata of leaves have varied tasks to fulfill. They guarantee the CO_2- and O_2-gas exchange between plant and atmosphere and cool the leaf by means of transpiration. Brownfields are predominantly dry sites. Thus, plant biomass production is reduced compared to sites with a favourable water supply, and the transpirational cooling is less effective. Water consumption has to be restricted by **stomatal closure**.

Additionally, a number of anatomical-morphological adjustments, typical of xerophytes, reduce the transpiration, such as a thick **cuticule** with high diffusional resistance, a dense layer of **hairiness**, or a **rolling** and **folding of leaves**. **Stomata in great density** or **situated in depressions** increase boundary layer resistance through the formation of a connected layer of water vapour above them.

Microphylly (miniature leaves) reduces the transpiring surface. **Mucilage cells** and **mucilage canals** delay desiccation and counteract the deformation of the cells caused by intense water loss. A raised concentration of potassium, sugars and nitrogen compounds acting as osmotica is energy consuming but increases the water potential of the cells, thus enabling an improved water uptake and simultaneously reducing water losses.

Dwarf growth reducing the distances for water transport, **enlargement of the conducting area** in the xylem and a dense leaf venation result in a more efficient water flux. A further strategy for coping with water shortage is an enlargement of the **root system**; extensive rooting plant species grow to the depth and enlarge their rooting system below the surface; intense rooting species condense their root system to gain water from as many soil pores as possible (comprehensive description i.e., LARCHER 2003, LÖSCH 2001).

An efficient adaptation strategy to drought stress is water storage through **succulence** (*Sedum sp.*, *Sempervivum sp.*) which is only successful if no competitive pressure of plants with overgrowing-strategy exists. Another advantage in coping with water stress is also the water-saving photosynthesis of **C4-plants** and the extreme water-saving metabolism of the (mostly) succulent **CAM-plants** (i.e., LAWLOR 1990).

These **arido-active strategies** are supplemented by **arido-tolerant strategies** within drought resistant plants (LÖSCH 2001). **Osmotic adjustment** is a modulatory and modificatory mechanism to increase desiccation tolerance. Water losses of the plant caused by dryness lead to a passive lowering of the osmotic water potential, which again deprives water from the cytosol. To protect the sensitive plasma structures from damaging by water loss, a variety of processes are activated at cellular level. Initially, drought stress leads to an increase in the content of **abscisic acid**, a fast signal that elicits stomatal closure. Furthermore, genes sensitive to abscisic acid initiate the synthesis of proteins such as **aquaporins** (transmembrane proteins facilitating water transport) and **protective proteins** (dehydrins) for membranes. Moreover, **enzymes** responsible for the expression of compatible solutes (i.e., carbohydrates, betaines, and amino acids such as proline) are synthesised in metabolically active compartments to counteract serious dehydration of the cytosol. Proteases dissimilate damaged proteins, and enzymes detoxifying reactive O_2-species are built (Lösch 2001, Schulze et al. 2002).

Finally, drought-sensitive plants use the **arido-passive strategy** (Lösch 2001) of spatial and temporal avoidance observed in therophytes and geophytes.

Wind

Wind is one of the variables influencing the effects of light, temperature and humidity. The boundary layer resistance above the leaves, which influences the diffusion rates of CO_2 and H_2O-vapour, is changed by the wind. Therefore wind exerts influence on the CO_2-supply of photosynthesis and transpiration by altering the layer of water vapour on the leaf surface. Transpiration rates do not necessarily need to be increased, since convective and advective transfer cool the leaves and adjust them to the air temperature. As a result, the water vapour pressure difference between leaf and air decreases and the transpiration rate drops. In winter, wind is disadvantageous, as cold wind affects leaves warmed up by solar radiation.
The plant's cuticle is reasonably resistant to chemical influences. However, it is vulnerable to mechanical influences to a certain degree and may be torn by the action of wind and particles transported by wind. Due to this fact the diffusion resistance of plant cells decreases, they loose water, thereby increasing osmotic potential, which again influences normal stomatal movement (Grace 1996).

Protective mechanisms against wind sensitivity are inter alia the development of **small leaves, solid leaf surfaces, dense epidermis** and **short petioles**. Mechanical stimultation is also perceived by chemical signals at cellular level, leading to anatomical modifications and influencing the distribution of assimilates (Grace 1996).

The importance of the wind factor for plants on temporary urban brownfields is its contribution to **seed dispersal**. Results of floristic investigations in railway areas showed that most species belong to the families Asteraceae, Poaceae, Brassicaceae and Fabaceae (Brandes 1983, Mattheis & Otte 1989, Mühlenbach 1979, Stalter 2004, Schinninger 2008). The reason for the high competitiveness and the wide dispersal of these families is above all the wind dispersal of seed. A competitive advantage for species of the Brassicaceae family is its allelopathic constituents. Fabaceae, again, have the ability of symbiotic nitrogen fixation.

3. Adaption strategies to anthropogenic factors

Human influences on brownfield vegetation

The vegetation of temporary habitats and in the urban landscape where humans exert formative influence, must adjust to changing ecological factors as well as deal with anthropogenic environmental factors. Examples of these factors are sulphur and nitrogen compounds, airborne dust, ions in the soil (such as nitrogen, sodium chloride and heavy metals), and synthetically produced substances, **xenobiotic compounds**, (i.e., insecticides and herbicides), all of which are greatly increased in their concentration by human activities. Other forms of manipulation include mowing, driving, walking and sealing.

Traffic, residential heating, small trades and industry are considered to be the causative agents of anthropogenic air pollution such as **sulphur dioxide** (SO_2) and **nitrogen oxide** (NO_x). Moreover, nitrogen oxides act as a precursor to photochemical oxidants. Further phytotoxic secondary air contaminations such as **ozone** (O_3) and **peroxyacetyl nitrate** (PAN) are produced by the combination of hydrocarbons emitted by motor vehicles and sunlight.

An adaptation strategy can be observed in the way that plants avoid the uptake of harmful gases through **stomatal closure** and **detoxify** penetrated gases. However, synergisms of negative effects may be triggered by the impact of various pollutant gases (Fuhrer 1996).

Aside from phytotoxic components, **dust** changes the absorption, reflection and transmission qualities of leaves. The temperature of plant organs may increase so much that burning causes damage (Eller & Brunner 1975). In the long term, thermal pollution of dusty plant organs has negative effects on productivity and water economy. The result of increased reflection of the visible spectral range is a supplementary reduction of photosynthetic active radiation and with that a further decrease in crop growth rate (Eller 1977, Sieghardt 1983). The negative effect of dust is worsened by the disturbance of stomatal movement (Flückinger et al. 1977, Godzik & Sassen 1978, Maier et al. 1979).

The classical 'metallophytes', with their avoidance and tolerance strategies in response to **heavy metals**, are not normally found in urban brownfields or in the other urban flora. The heavy-metal concentration in urban soils may be very high, but plants may suffer from an undersupply in essential heavy metals in neutral to basic urban soils (Braun 1990), as heavy metals are frequently not plant-available in basic milieu – contrary to acid ground conditions. Adaptations to iron (Fe) deficiency, for example, are visible in an **acidification of the rhizosphere** through ATPase-controlled release of protons by roots, or through the induction of Fe^{3+}-chelate reductase. Grasses mobilise iron by excreting chelating substances. A surplus of heavy metals has a toxic effect. If heavy metals are not bound to the **apoplast** but get to the cytosol, they deactivate enzymes; proof that heavy metals are extracted from the plasma by transport to the **vacuole** and that the cell reacts via synthesis of **phytochelatins** (Ernst 1996). However, heavy metals can also lead to changes in the soil indirectly causing plant damage (Maier et al. 1981). Another concern about heavy metals is their accumulation in soil and plants, resulting in a lengthy presence in the food chain.

Sodium chloride (NaCl) is frequently used as a thawing agent on wintry roads and pavement, which mainly affects the green areas alongside roads. Yet the toxicity of de-icing salt is not obvious in the herbaceous plant stands of the brownfields. However, toxicity provides subjects for discussion about environmental quality among woody species, above all among trees. Peripheral woods of urban brownfields may exhibit classic damage symptoms: yellowing and foliar necrosis.
Through transpiration, chloride accumulates in the leaf, apoplastic present chloride extracts water from cells and leads to a strained water balance (i.e., Albert et al. 1989), and symplastic enriched chloride leads to metabolic disorders (i.e., Griessler 1985, Maier et al. 1989). Moreover, the stress situation is aggravated by changes in the soil caused by salt. The soil water becomes osmotic effective, the pH-value changes, a decomposition of colloids condenses the soil, and the capillary bonding of water increases (i.e., Glatzel & Krapfenbauer 1975). Unfavourable soil conditions are transmitted to the activity of soil microorganisms, and the mineralisation of nitrogen is lowered (Albert 1987).
Morphologic-anatomical adjustments to salt – as realised within halophytes – were not developed by urban plants. The **stress response** takes place via cellular mechanisms, as it does for drought stress.

The use of urban brownfields as urban recreation areas of high experiential value (Kinzl 2007), as children's play area, as dog-exercising area but also use as car park cause eutrophication and soil compaction.

Eutrophication leads to **nitrophilous plant communities**, such as *Chenopodium*- and *Urtica*-stands. Eutrophication is also an aesthetic and hygienic problem, due to the amount of dog excrement and pigeon droppings. For example, on Vienna`s streets more than 30 tons of dog excrement accumulates daily (Stadt Wien MA 48 2005). The annual amount of pigeon droppings in Lucerne is about 80 tons (Stadt Luzern 2003).

Soil compaction excerts a critical influence on plant growth by limiting water uptake and gas exchange, as well as by the limited root zone in green areas situated next to roads.
Driving, trampling and the effects of thawing salt compact the soil (above all micro- and macropores are affected) and thus deplete the amount of plant-available water (Meyer 1982). Adaptations are visible in **plants of trampled grounds** inter alia with *Plantago maior* and *Polygonum aviculare*, which are specialised to these particular disturbances. After people were allowed to walk on lawns in Vienna it was observed that the soil fauna in the densified topsoil was decimated, and that the vegetation began to change to plant communities adjusted to trampling (Kühnelt 1989). On Viennese brownfields *Polygonum aviculare* is among the five most common plants of the herbaceous layer (Punz et al. 1998).
Soil compaction also leads to hypoxia in the soil. Leaking gas pipelines also may contribute to the replacement of soil air (Wittig 1998). Methane bacteria that use natural gas as a carbon source additionally consume oxygen (Ruge 1982). Moreover, the negative redox potential under hypoxia mobilises heavy metals in the soil. As a rule, the primary root system of terrestrial plants does not tolerate oxygen deficiency. Hypoxia leads to, amongst other problems, impairment of energy metabolism through switching from dissimilation to fermentation (comprehensive description i.e., Schulze et al. 2002). Ethylene, acting as a signal, elicits the formation of **aerenchyma** in marsh plants: *Phragmites australis*, which can be found in water as well as on urban brownfields, has an aeration tissue with significant convective gas throughputs (Armstrong & Armstrong 1990) and is thus adapted to soil conditions low in oxygen. Compacted soils are less densely-rooted; plants extend their roots through earthworm tunnels, in the hollow spaces of rotten roots (Lösch 2001) or restrict their root system to near the ground's surface (Larcher 2003).

Different strategies enable plants to adapt well to the continuous infrastructural changes of the 'ecosystem town' and the prevailing environmental conditions. Plant species and individuals are expected to respond differently to their environment (Bazzaz 1996), whereas they are to a certain degree 'flexible' and react with plasticity to their particular surroundings. Warmth- and nutrient-loving species, species that are able to tolerate disturbances, as well as species that are drought resistant, wind dispersed and short-lived, are especially supported in urban environments.

The effects of shrinking cities and brownfield integration present a challenge to the economic-social oriented concept of sustainable urban planning (Maier 1992). The town as a habitat for humans, plants and animals needs those green areas which have important ecological, social and economical potential (Gebhard 1998, Kinzl et al. 2006, Kinzl 2007).

References

Albert, R. (1987): Alleen in Wien – Zustandserhebung und Ergebnisse von Bodensanierungsmaßnahmen. Magistrat der Stadt Wien. Wien.

Albert, R., Schestag-Schafellner, Ch. & V. Trockner-Di Costigliolie (1989): Auswirkung innerörtlicher Stressfaktoren auf baumphysiologische Parameter. Kali-Briefe 19: 743-765.

Armstrong, J. & W. Armstrong (1990): Light enhanced convective throughflow increases oxygenation in rhizomes and rhizosphere of *Phragmites australis* (Cav.) Trin. ex Steud. New Phytol. 114: 121-128.

Bais, H. P., Vepachedu, R., Gilroy, S., Callaway, R. M. & J. M. Vivanco (2003): Allelopathy and exotic plant invasion: From molecules and genes to species interactions. Science 301: 1377-1380.

Bazzaz, F. A. (1986): Life history of colonizing plants: some demographic, genetic, and physiological features. In: Mooney, H. A. & J. A. Drake (eds.): Ecology of Biological Invasions of North America and Hawaii. Berlin: 96-110.

Bazzaz, F. A. (1996): Plants in Changing Environments. Cambridge.

Bohlmann, F., Burkhardt, T. & C. Zedro (1973): Naturally occurring acetylenes. New York.

Bornkamm, R. (1986): Ruderal succession starting at different seasons. Acta Soc. Bot. Poloniae 55: 403-419.

Bradshaw, A. D. (1984): Ecological significance of genetic variation between populations. In: Dirzo, R. & J. Sarukhán (eds.): Perspectives on Plant Population Ecology. Sinauer Associates Inc. Sunderland: 213-228.

Brandes, D. (1983): Flora und Vegetation der Bahnhöfe Mitteleuropas. Phytocoenologia 11: 31-115.

Brandes, D. & D. Zacharias (1990): Korrelation zwischen Artenzahlen und Flächengrößen von isolierten Habitaten, dargestellt an Kartierungsprojekten aus dem Bereich der Regionalstelle 10 B. Floristische Rundbriefe 23: 141-149.

BRAUN, CH. (1990): Zustand der Wiener Stadtbäume. Vergleichende Kronenzustandserhebung und Untersuchung des Mineralstoffhaushaltes. Dissertation Universität Wien. Wien.

CALLAWAY, R.M. (1995): Positive interactions among plants. Botanical Review 61: 306-349.

CALLAWAY, R. M. & L. R. WALKER (1997): Competition and facilitation: A synthetic approach to interactions in plant communities. Ecology 78: 1958-1965.

CALLAWAY, R. M. & J. L. HIERO (2006): Resistance and susceptibility pf plant communities to invasion: Revisiting Rabotonov`s ideas about community homeostasis. In: REIGOSA, M. J., PEDROL, N. & L. GONZALEZ (eds.): Allelopathy: A physiological Process with ecological implications. Dordrecht: 395-414.

CARDINALE, B. J., PALMER, M. A. & S. L. COLLINS (2002): Species diversity enhances ecosystem functioning through interspecific facilitation. Nature 415: 426-429.

CASPER, B. B. & R. B. JACKSON (1997): Plant competition underground. Annual Review of Ecology and Systematics 28: 545-570.

DE FEO, V., DE MARTINO, L., QUARANTA, E. & C. PIZZA (2003): Isolation of Phytotoxic Compounds from Tree-of-Heaven (*Ailanthus altissima Swingle*). J. Agric. Food Chem. 51: 1177-1180.

ELLER, B. M. (1977): Road dust induced increase of leaf temperature. Environ. Pollut. 13: 99-107.

ELLER, B. M. & U. BRUNNER (1975): Der Einfluss von Straßenstaub auf die Strahlungsabsorption durch Blätter. Arch. Met. Geoph. Biokl. B23: 137-146.

ERNST, W. H. O. (1996): Schwermetalle. In: BRUNOLD, CH., RÜEGSEGGER, A. & R. BRÄNDLE (eds.): Stress bei Pflanzen. Bern: 191-219.

ESCHRICH, W. (1995): Funktionelle Pflanzenanatomie. Berlin.

ESSL, F. & W. RABITSCH (2002): Neobiota in Österreich. Umweltbundesamt, Wien.

FLÜCKIGER, W., FLÜCKIGER-KELLER, H., OERTLI, J. J. & R. GUGGENHEIM (1977): Verschmutzung von Blatt- und Nadeloberflächen im Nahbereich einer Autobahn und deren Einfluss auf den stomatören Diffussionswiderstand. J. Forest Path. 7: 358-364.

FUHRER, J. (1996): Gasförmige Luftschadstoffe. In: BRUNOLD, CH., RÜEGSEGGER, A. & R. BRÄNDLE (eds.): Stress bei Pflanzen. Bern: 191-219.

GEBHARD, U. (1998): Stadtnatur und psychische Entwicklung. In: SUKOPP, H. & R. WITTIG (eds.): Stadtökologie. Ein Fachbuch für Studium und Praxis. Stuttgart: 105-124.

GILBERT, O. L. (1994): Städtische Ökosysteme. Radebeul.

GLATZEL, G. & A. KRAPFENBAUER (1975): Streusalzschäden am Baumbestand der Straßen in Wien. Studie d. Inst. f. forstl. Standortsforschung d. Hochsch. f. Bodenkultur. Wien.

Godefroid, S. & N. Koedam (2007): Urban plant species patterns are highly driven by density and function of built-up area. Landscape Ecology. DOI 10.1007/s10980-007-9102-x.

Godzik, S. & M. M. Sassen (1978): A scanning electron microscope examination of *Aesculus hippocastanum* L. leaves from control and air-polluted areas. Environ. Pollut. 17: 13-18.

Goldberg, D. E. & A. M. Barton (1992): Patterns and consequences of interspecific competition in natural communities: a review of field experiments with plants. American Naturalist 139: 771-801.

Grace, J. (1996): Wind. In: Brunold, Ch., Rüegsegger, A. & R. Brändle (eds.): Stress bei Pflanzen. Bern: 173-188.

Griessler, B. (1985): Nachweis der Einzel- und Kombinationswirkungen von Blei und Streusalz auf Proteine und unspezifische Enzyme von *Zea mays* und *Pisum sativum* mit Hilfe der Gelelektrophorese. Dissertation Universität Wien. Wien.

Grime, J. P. (1973): Competitive exclusion in herbaceous vegetation. Nature 242: 344-347.

Grime, J. P. (1977): Evidence for the existence of three primary strategies in plants and its relevance to ecological and evolutionary theory. American Naturalist 111: 1169-1194.

Grime, J. P. (2001): Plant strategies, vegetation processes and ecosystem properties. Chichester.

Hegi, G. (1979): Illustrierte Flora von Mitteleuropa. Berlin.

Heisey, R. M. (1990): Evidence for allelopathy by tree-of-heaven (*Ailanthus altissima*). J. Chem. Ecol. 16: 2039-2055.

Holmgren, M., Scheffer, M. & M. A. Huston (1997): The interplay of facilitation and competition in plant communities. Ecology 78: 1966-1965.

Huang, B., (ed.) (2006): Plant-environment interaction. Boca Raton.

Joshi, J., Schmid, B., Caldeira, M. C., Dimitrakopoulos, P. G., Good, J., Harris, R., Hector, A., Huss-Danell, K., Jumpponen, A., Minns, A., Mulder, C. P. H., Pereira, J. S., Prinz, A., Scherer-Lorenzen, M., Siamantziouras, A.-S. D., Terry, A. C., Troumbis, A. Y. & J. H. Lawton (2001): Local adaptation enhances performance of common plant species. Ecology Letters 4: 536-544.

Kaussmann, B. & U. Schiewer (1989): Funktionelle Morphologie und Anatomie der Pflanzen. Stuttgart.

Kinzl, C., Punz, W. & R. Maier (2006): Ein Indikatorenschlüssel zur Bewertung von Brachflächen in Wien. Beitr. z. Naturkunde OÖ 16: 631. 75 Jahre Botanische Arbeitsgemeinschaft. Biologiezentrum OÖ Landesmus. Linz/Austria.

Kinzl, C. (2007): Freilandökologische Untersuchungen zur Erstellung eines Indikatoren- und Bewertungsschlüssels von Brachflächen in Wien. Diplomarbeit Universität Wien. Wien.

KOWARIK, I. (1986): Vegetationsentwicklung auf innerstädtischen Brachflächen – Beispiele aus Berlin (West). Tuexenia 6: 75-98.

KOWARIK, I. (1995): On the role of alien species in urban flora and vegetation. In: PYSEK, P., PRACH, K., REMANEK, M. & M. WADE (eds.): Plant invasions: general aspects and special problems. Amsterdam: 85-103.

KOWARIK, I. & R. BÖCKER (1984): Zur Verbreitung, Vergesellschaftung und Einbürgerung des Götterbaumes (*Ailanthus altissima* [Mill.] Swingle) in Mitteleuropa. Tuexenia 4: 9-29.

KOWARIK, I. & A. LANGER (1994): Vegetation einer Berliner Eisenbahnfläche (Schöneberger Südgelände) im vierten Jahrzehnt der Sukzession. Verh. Bot. Ver. Berlin Brandenburg 127: 5-43.

KOWARIK, I. & I. SÄUMEL (2007): Biological flora of Central Europe: *Ailanthus altissima* (Mill.) Swingle. Perspect. Plant Ecol., Evol. Systematics 8(4): 207-237.

KRATZER, P. (1956): Das Stadtklima. Braunschweig.

KÜHNELT, W. (1989): Characteristics and development of urban soil fauna. MAB-Mitteilungen 30: 57-70.

KUTTLER, W. (1998): Stadtklima. In: SUKOPP, H. & R. WITTIG (eds.): Stadtökologie. Ein Fachbuch für Studium und Praxis. Stuttgart: 125-167.

LANGENHEIM, J. H. (1994): Higher-plant terpenoids – A phytocentric overview of their ecological roles. J. Chem. Ecol. 20: 1223-1280.

LARCHER, W. (2003): Physiological plant ecology. Ecophysiology and stress physiology of functional groups. Berlin.

LAWLOR, D. W. (1990): Photosynthese. Stoffwechsel-Kontrolle-Physiologie. Stuttgart.

LEVITT, J. (1980): Responses of plants to environmental stresses. Vol I+II. Orlando.

LÖSCH, R. (2001): Wasserhaushalt der Pflanzen. Wiebelsheim.

MACARTHUR, R. (1962): Some generalized theorems of natural selection. Proc. Natl. Acad. Sci. USA 48: 1893-1897.

MACARTHUR, R. & E. O. WILSON (1967): The theory of island biogeography. Princeton.

MAIER, R. (1992): Pflanzen im Lebensraum Stadt. In: FRANZ, H. (ed.): Die Störung der ökologischen Ordnung in den Kulturlandschaften. Veröff. d. Komm. f. Humanökologie 3. Österr. Akad. Wiss. Wien: 145-181.

MAIER, R., ALTGAYER, M., PUNZ, W., RAMMER, CH., SCHINNINGER, R., SIEGHARDT, H., SLAD, H. & CH. WINTER (1979): Wasserhaushalt und Produktivität staubbelasteter Pflanzen in der Umgebung einer Zementfabrik in Kärnten. Carinthia II 169/89: 167-193.

MAIER, R., MAIER, F. & G. MAIER (1981): Aktivität TTC-reduzierender Enzyme im Boden und Pflanze unter dem Einfluss von Blei im Boden. Ber. Dtsch. Bot. Ges. 94: 709-718.

Maier, R., Wiltschegg, U. & W. Rücker (1989): Effects of salt and heavy metals on growth and isoelectric peroxidase pattern in callus cultures of horse-chestnut. Electrophoresis Forum 89: 570-574.

Mattheis, A. & A. Otte (1989): Die Vegetation der Bahnhöfe im Raum München-Mühldorf-Rosenheim. Ber. ANL 13: 77-143.

Meyer, F. H. (1982): Lebensbedingungen der Straßenbäume. In: Meyer, F. H. (ed.): Bäume in der Stadt. Stuttgart: 84-133.

Mooney, H. M., Winner, W. E. & E. J. Pell, (eds.) (1991): Responses of plants to multiple stresses. San Diego.

Mühlenbach, V. (1979): Contributions to the synanthropic (adventive) flora of the railroads in St. Louis, Missouri, USA. Ann. Miss. Bot. Garden 66: 1-108.

Odum, E. P. (1969): The strategy of ecosystem development. Science 169: 262-270.

Pianka, E. R. (1970): On r- and K-selection. American Naturalist 104: 592-597.

Prasse, R. & M. Ristow (1995): Die Gefäßpflanzen einer Berliner Güterbahnhofsfläche (Schöneberger Südgelände) im vierten Jahrzehnt der Sukzession. Verh. Bot. Ver. Berlin-Brandenburg 128: 165-192.

Punz, W., Aigner, B., Schimpl, C., Pietsch, G., Schosmeier, E. & R. Maier (1998): Stadtbrachen in Wien. Verh. Zool.-Bot. Ges. Österr. 135: 171-184.

Punz, W., Kober, M., Armeanu, K., Kugler, R., Engenhart, M., Schinninger, I., Sieghardt, H. & R. Maier (2004): Beiträge zur Ökophysiologie von *Ailanthus altissima* im Raum Wien. Verh. Zool.-Bot. Ges. Österr. 141: 1-11.

Pysek, P. & A. Pysek (1991): Succession in urban habitats: an analysis of phytosociological data. Preslia 63: 125-138.

Raunkiær, C. (1910): Statistik der Lebensformen als Grundlage für die biologische Pflanzengeographie. Beih. Biol. Cbl. 27: 171-206d.

Raunkiær, C. (1934): The life form of plants and statistical plant geography. Oxford.

Reigosa, M. J., Pedrol, N. & L. Gonzalez, (eds.) (2006): Allelopathy: A physiological Process with ecological implications. Dordrecht.

Rice, E. L. (1977): Some roles of allelopathic compounds in plant communities. Biochem. Syst. Ecol. 5: 201-206.

Richter, G. (1996): Biochemie der Pflanze. Stuttgart.

Ruge, U. (1982): Physiologische Schäden durch Umweltfaktoren. In: Meyer, F. H. (ed.): Bäume in der Stadt. Stuttgart: 134-198.

Sieghardt, H. (1983): Beeinflussung der spektralen Absorptionseigenschaften der Blätter von *Hedera helix* durch Staubimmissionen. Phyton 23: 177-184.

Schinninger, I. (2005): Ökologische Charakterisierung brachliegender Bahnareale in Wien – Vegetation und Wasserhaushalt. Dissertation Universität Wien. Wien.

Schinninger, I. (2008): Die Bedeutung brachliegender Bahnareale als Lebensraum für Pflanzen am Beispiel der Stadt Wien. Braunschweiger Geobotanische Arbeiten 9: 393-404.

Schinninger, I., Maier, R. & W. Punz (2002): Der stillgelegte Frachtenbahnhof Wien-Nord. Standortbedingungen und ökologische Charakteristik der Gefäßpflanzen einer Bahnbrache. Verh. Zool.-Bot. Ges. Österr. 139: 1-10.

Schinninger, I., Maier, R. & W. Punz (2003): Der stillgelegte Frachtenbahnhof Wien-Nord. Stadtökologische und ökophysiologische Aspekte. IÖR-Schriften 39: 85-95.

Schlee, D. (1992): Ökologische Biochemie. Jena.

Schippers, P., J. M. Groenendael, J. M., Vleeshouwers, L. M. & R. Hunt (2001): Herbaceous plant strategies in disturbed habitats. Oikos 95: 198-210.

Schoener, T. W. (1983): Field experiments on interspecific competition. American Naturalist 122: 240-285.

Schulze, E.-D., Beck, E. & K. Müller-Hohenstein (2002): Pflanzenökologie. Heidelberg.

Stadt Luzern (2003): "Stadttauben Luzern" – das Projekt für eine kleine und gesunde Taubenpopulation. Luzern.

Stadt Wien, MA 48 (2005): Littering in Wien. Powerpointvortrag, 7.6.2005, Magistratsabteilung 48. Wien.

Stalter, R. (2004): The flora on the High Line, New York City, New York. J. Torrey Bot. Soc. 131: 387-393.

Sukopp, H., Blume, H. P., Chinnow, D., Kunick, W., Runge, M. & F. Zacharias (1974): Ökologische Charakteristik von Grossstädten, besonders anthropogene Veränderungen von Klima, Boden und Vegetation. Zeitschr. Techn. Univ. Berlin 6: 469-488.

Tilman, D. (1982): Resource competition and community structure. Princeton.

Tilman, D. (1988): Plant strategies and the dynamics and structure of plant communities. Princeton.

Tilman, D. (1994): Competition and biodiversity in spatially structured habitats. Ecology 75: 2-16.

Trepl, L. (1994): Zur Theorie urbaner Biozönosen. Einige Hypothesen und Forschungsfragen. Geobot. Kolloq. 11: 17-32.

Troll, W. & K. Höhn (1973): Allgemeine Botanik. Stuttgart.

Whittaker, R. H. (1965): Dominance and diversity in land plant communities. Science 147: 250-260.

Wittig, R. (1998): Flora und Vegetation. In: Sukopp, H. & R. Wittig (eds.): Stadtökologie. Ein Fachbuch für Studium und Praxis. Stuttgart: 219-265.

URBAN FORM AND BIODIVERSITY

Peter Werner

1. Introduction

It is now largely accepted that there is a connection between urban form, as defined by the parameters size, density, and compactness, and the ecological impacts of cities. It is therefore not surprising that there has been intensive scientific and political discourse on urban form and sustainable urban development in recent years (JENKS & DEMPSEY 2005, WILLIAMS et al. 2003). The discussion centres on the compact city (JENKS et al. 1996, JENKS & BURGESS 2000). Especially in Europe, the historically grown city defined by clear development limits and a densely developed centre is regarded as a model. Many town, regional, and environmental planners dealing with sustainable urban development regard the compact city as the only sustainable urban form, a form to strive for and to defend.

In recent years this discussion has gained a new dimension, i.e. that of shrinking cities. Thus the compact city as a planning ideal is no longer just threatened by suburbanization and urban sprawl but now also by shrinking processes. This article will not discuss the compact city as a model or its importance for sustainable urban development but will present research results which provide indications of the connection between urban form and biodiversity. However, the compact city and its dissolution by different types of shrinking such as deconcentration, perforation, and reduction is an important starting point for this analysis.

2. Data and methods

The results presented here are derived from a study which aimed to test whether it is possible to describe and evaluate connections between urban form and biodiversity using available biological data and publicly accessible municipal statistical data. Data accessibility was thus a key concern; in this context it means that the data are available at the Institute for Housing and Environment and are freely useable, such as data maintained by the institute on behalf of the regional state (Land) of Hesse.

Data compiled in the context of urban biotope mapping were used to describe biodiversity. Urban biotope maps have been produced for more than 220 German towns and cities (Fig. 1). Almost every city with more than 100,000 inhabitants has produced such a map. For this test study, data from 21 towns and cities were evaluated for which the author has complete full flora lists at hand from the biotope mapping schemes. Full flora lists are basically the only biological data available for every town or city which has completed biotope mapping schemes. Faunal data compiled

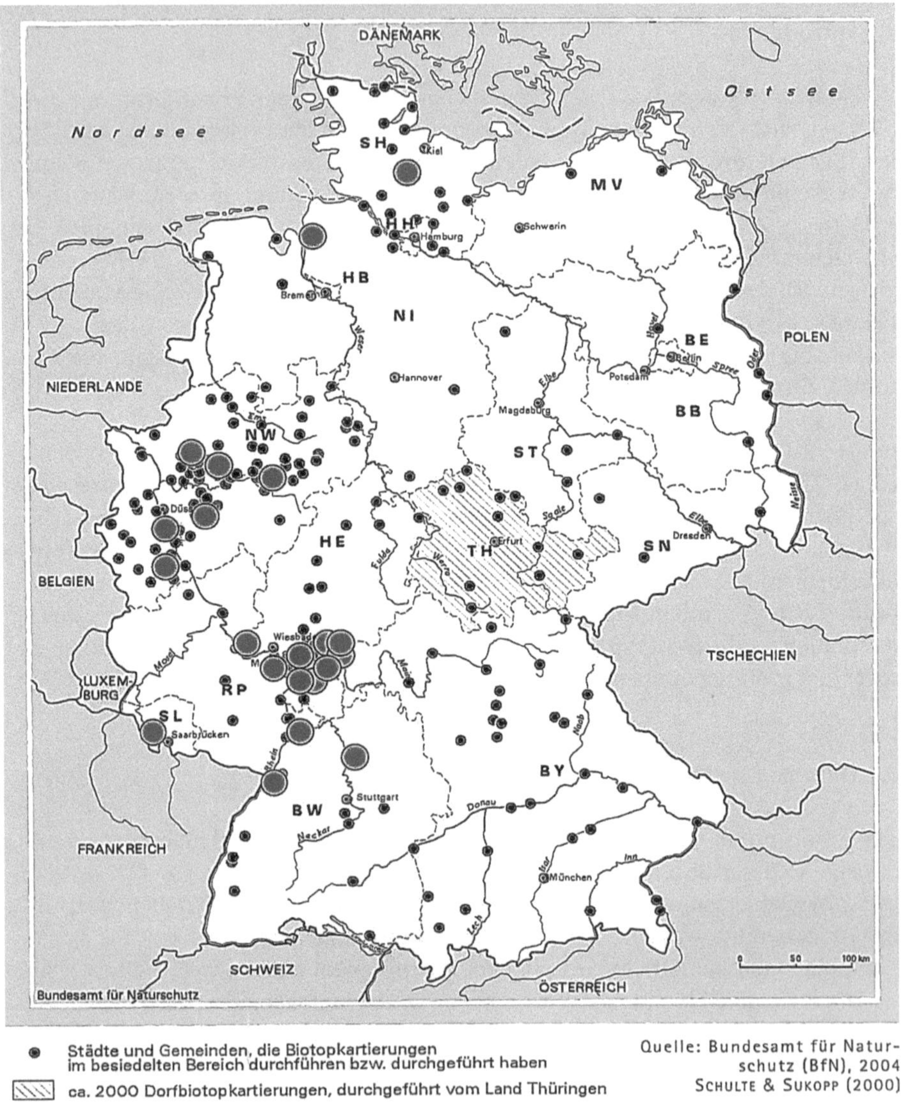

Fig. 1: Distribution of urban biotope mappings in Germany. The big grey circles show the test cities and towns

under these schemes, in contrast, are much more heterogeneous and not as well suited for comparisons.

The flora lists of the test towns were evaluated with respect to the following characteristics of the individual plant species:

- naturalization status;
- botanical life form;
- ecological requirements (indicator values after ELLENBERG);
- phytosociological classification.

For the description of urban form, characteristics were selected which are descriptive of town size, town outline, urban density, heterogeneity of land uses (ALBERTI 2000) and the embedding of settlements into land uses in the cultural landscape. For the 21 sample towns the following data were extracted from the municipal statistics:

- area of the municipality;
- settlement area;
- extent of the different types of land uses;
- number of inhabitants;
- number of buildings and apartments.

Only the shape of the urban form had to be deduced from maps and aerial photographs and converted into stylized maps as no suitable indicators could be derived from the municipal statistics.

Biodiversity and sustainable urban forms

It has been shown many times that urban areas host a large number of plant and animal species and that more species per square kilometre can be found in cities and towns than in agricultural or forestry areas (SUKOPP & WITTIG 1998, REICHHOLF 2007). It is also well known that there is a direct positive correlation between the size of a city respectively town and the number of plant species it hosts. This correlation is also evident in the sample cities and towns (Fig. 2). The fundamental reason for this correlation is that habitat diversity increases with the size of urban areas. It is also known that from an urban's outskirts to its centre there is a gradient of increasing intensity of use and decreasing number of plant and animal species. Both urbanization and increasing intensity of urban land use alter species composition. Finally, it is known that different types of urban land use have characteristic suites of species.

Not only plant species can be used to show that with increasing urbanization the proportion of hemerochorous species, i.e. the share of archaeophytes and neophytes, increases, often at the expense of native species (OLDEN et al. 2006). Amongst plant

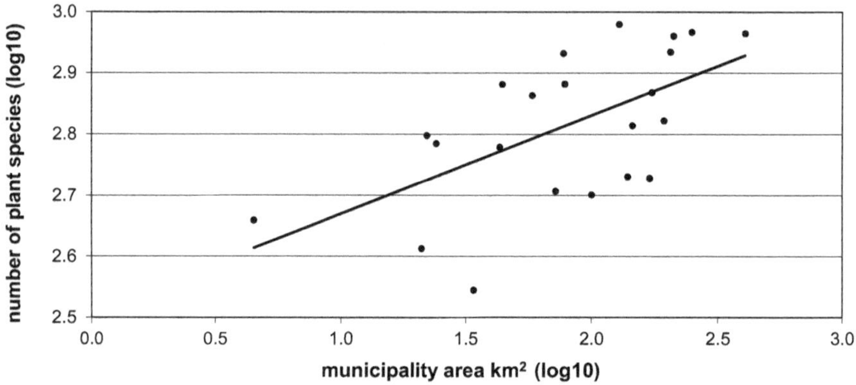

Fig. 2: Species-area-relationship of the 21 tested cities and towns ($r^2 = 0.3615$, $p \leq 0.003$)

species, the number of species requiring higher levels of nitrogen, light, and heat increases (Wittig 2002).

Even though it is relatively well known how flora and vegetation are influenced by cities and urbanization, little is known about the impacts of different urban settlement structures on the composition of flora and vegetation, i.e. it is not clear how the relationship between a compact city and its species diversity compares to low-density settlement structures. The indicators listed above can be used to estimate the impact of different urban structures on biodiversity at the local level. Other data must be used at the regional level (see e.g. Kühn & Klotz 2006).

Shrinking cities and urban form

In order to be able to assess potential impacts of shrinking cities on biodiversity it is necessary to record the structural changes caused by the shrinking process. Three typical types of processes of change can be used to describe the shrinking process (Fig. 3):

- Decreasing intensity of use
  The average number of people per unit area, both inhabitants and employees, as well as the share of multi-storey buildings is declining.
- Perforation
  Individual built-on sites become transformed into brownfield or unused sites in a disorganized fashion and across the entire municipal area. The extent and distribution of this transformation can vary significantly.

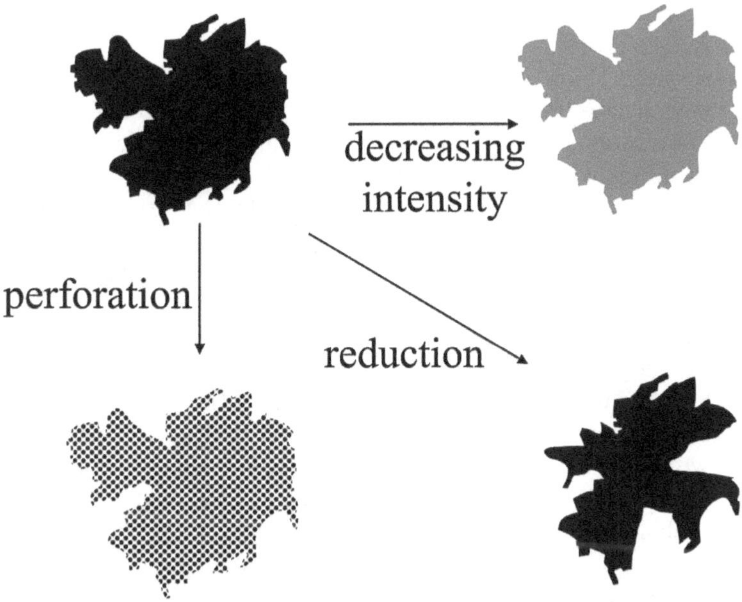

Fig. 3: Decreasing intensity, perforation and reduction are the three types of shrinking processes of cities

- Reduction
  The removal of buildings and infrastructure from larger areas ('dismantling') changes the original urban shape from the edge inwards and open spaces penetrate into formerly built-up areas.

The first two processes can be summarized as deconcentration. All three processes entail a loss of compactness.

The 21 selected sample towns were sorted and grouped on the basis of the characteristics 'intensity of use' (inhabitants per unit area and ratio of multi-storey to low buildings), 'perforation' (ratio of open spaces to built-up spaces) and 'reduction' (ratio of compact town shapes to 'ragged-edge' town shapes) with the aim of simulating potential effects of shrinking processes by comparing the sample towns. Regression analyses were used as much as possible to evaluate the effects.

3. Results

The selected sample towns vary in terms of

- size of town, from 21 km² (Mühlheim/M.) up to 405 km² (Cologne);
- size of settlement area, from 6 km² (Mühlheim/M.) up to 233 km² (Cologne) and from 18 % (Arnsberg) to 65 % (Essen) of total area;
- extent of the different types of land uses:
  - built-up sites comprise between 52 % (Wiesbaden) and 70 % (Dietzenbach) of the settlement area,
  - designated green spaces comprise between 2 % (Dietzenbach) and 16 % (Wiesbaden) of the settlement area;
- number of inhabitants, from 24,000 inhabitants (Mühlheim/M.) up to 914,000 inhabitants (Cologne) and in terms of population density from 1,600 inhabitants/km² (Saarlouis) up to 4,800 inhabitants/km² (Wuppertal);
- development intensity expressed as the ratio of the number of housing units in buildings with more than three units to the number in buildings with one to two units, with values of 0.9 (Mörfelden-Walldorf) up to 6.6 (Frankfurt/M.). Figures are not available for all cities.

Biodiversity and decreasing intensity of use

In order to ascertain the connection between the intensity of use and biodiversity, the number of inhabitants per unit of settlement area and the number of inhabitants per unit of built-up area were put in relation to the floristic indicators. In the land-use statistics settlement areas include four land uses, i.e. buildings and open spaces, transport infrastructure, recreational areas, and cemeteries. The 'built-up area' equates to the statistical category of buildings and open spaces.

The analyses show that the share of neophytes in the flora is the parameter which correlates strongest with the indicators of land-use intensity as listed above. Similar findings were made in gradient analyses, analyses of different settlement types, and comparisons of urban structural types (Pysek 1998). The absolute number of plant species or differentiations by life-form or ecological requirements were not found to be significant indicators in the comparisons.

However, a comparison of cities and towns belonging to the same size class (Fig. 4) – the 21 cities and towns had been grouped into four size classes – did not yield any indication that reduced intensity of use resulted in a change in the share of neophytes. It appears that town size is the attribute which masks the other variables. Numbers of inhabitants are normally closely correlated with the size of settlement and built-up areas. Hence, data from studies which exclusively relate to numbers

Tab. 1:Core characteristics of the test cities and towns. The floristic data of the biotope mappings are from the 1980ies and 1990ies, the data of the local statistics usually correspond with this period

city	population	munici-pality area in km²	settlement area in km²	construction area in km²	number of plant species	neophytes %	archaeo-phytes %	therophytes %	indicator value N
Arnsberg	74,000	193	35	23	664	9.9	12.7	20.5	5.16
Bochum	383,000	145	98	62	457	17.7	14.0	25.2	5.58
Bremerhaven	117,000	78	36	22	762	13.7	16.7	20.1	5.43
Darmstadt	134,000	128	41	24	955	15.0	12.4	24.8	4.77
Dietzenbach	27,000	22	7	5	625	12.2	14.6	25.1	5.08
Essen	620,000	210	137	81	909	18.2	12.8	27.4	5.36
Euskirchen	45,000	140	27	16	544	15.3	19.3	29.6	5.55
Frankfurt	645,000	248	137	77	926	18.0	13.4	28.2	5.24
Hanau	85,000	77	29	18	854	16.3	13.2	25.1	5.24
Heilbronn	121,000	100	33	20	502	14.4	11.8	21.8	5.18
Karlsruhe	268,000	173	77	46	738	18.7	15.4	27.3	5.20
Köln	914,000	405	237	131	922	18.3	12.4	25.3	5.22
Mannheim	296,000	145	79	48	649	17.4	15.6	31.1	5.38
Mörfeld.-Walld.	30,000	44	9	6	761	16.8	15.0	24.7	5.14
Mühlheim	24,000	21	6	4	410	10.5	14.9	17.1	4.93
Neu-Isenburg	35,000	24	8	5	604	14.9	13.7	24.2	5.26
Neumünster	83,000	72	33	22	509	11.8	10.6	20.0	5.14
Rüsselsheim	57,000	58	17	10	730	15.0	16.1	25.4	5.29
Saarlouis	38,000	43	23	14	601	14.3	17.1	27.6	5.21
Wiesbaden	267,000	204	77	40	863	14.0	12.2	21.2	5.52
Wuppertac	38,000	170	80	48	534	16.1	12.0	17.8	5.39

	Karlsruhe	Mannheim
settlement area	77 km²	80 km²
population	268,000	385,000
pers./km²	3,481	4,842

Fig. 4: Two cities representing different intensities of urban land use

of inhabitants and do not consider the size of settlement areas at the same time are primarily influenced by species-area-relationships or by the correlation between area size and habitat diversity.

Biodiversity and perforation

It would be possible to study potential impacts of perforation by comparing degrees of mixture of open spaces and built-up spaces. However, municipal statistics do not offer any data on the spatial distribution and the individual sizes of public green spaces or on the extent of private green spaces. Therefore only the overall ratio of public green spaces to settlement areas and built-up spaces respectively can serve as indicators.

Here the comparison between cities and towns in equal size classes shows that the share of plant species with higher nitrogen requirements increases with an increase in the share of green spaces. A possible explanation may be that the manner in which green spaces are maintained by the towns' parks departments leads to this effect. The reverse implication would be that if in shrinking towns a major share of the released sites were outside the sphere of responsibility of parks management, this effect should no longer be evident.

The present results should only be regarded as indications as the number of towns in each size class is not large enough to yield statistically relevant results.

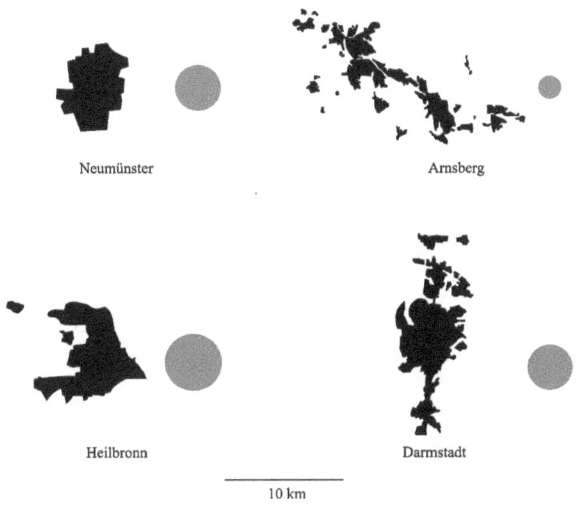

Fig. 5: Comparison of the compactness by maximum circle areas for four cities of the same size group

Biodiversity and reduction

Many planners view the loss of compactness of a city also as a loss of environmental sustainability.

Assessing the impact of a loss of compactness in urban form on the vegetation of a town requires completely different data than those mentioned so far. In conservation circles the SLOSS concept – single large or several small – has been widely and intensively discussed. It is generally held that large, contiguous areas and areas with a low edge-area ratio are superior as nature reserves compared to small, separated or narrow areas (Wilson & Willis 1975). This concept of more or less advantageous spatial structures can also be applied to urban areas in so far as large contiguous urban areas represent more effects of urbanization than small, dispersed settlement structures.

The shape forms of the sample cities were transferred into simplified black and white images so as to allow for the essential shapes of the settlement areas to be made visible and amenable to evaluation. Green spaces framed by built-up sites were assigned to settlement areas. Linear green spaces reaching from the undeveloped periphery into the towns' interior were however considered as exterior lands and not included in the settlement area. Fig. 5 shows some examples of urban forms of varying degrees of compactness.

Tab. 2: Compactness indices of four cities of the same size group

city	population	max. circle area km²
Arnsberg	74,000	2.5
Darmstadt	134,000	7.1
Neumünster	83,000	10.2
Heilbronn	121,000	15.9

To assess the urban forms encountered in the study a very simple index of compactness was chosen. Since a circle represents the most compact shape, the biggest possible circular area that could fit into the shape of the contiguous settlement area was used to define the individual town's degree of compactness (Tab. 2).

This figure was then related to the floristic data. Again, the share of neophytes was the parameter which correlated significantly with the index of compactness. The shares of archaeophytes or therophytes however – the latter are also considered an indicator of urbanization – did not show any considerable correlation to the index of compactness.

It was mentioned above that in town planning compactness is also considered an indicator of urbanness. In the introductory remarks it was shown that the share of hemerochores – the sum of archaeophytes and neophytes – has been regarded as a measure of urbanization at least since FALINSKI 1971. The results presented here show that the share of neophytes alone is a useful indicator of urbanization.

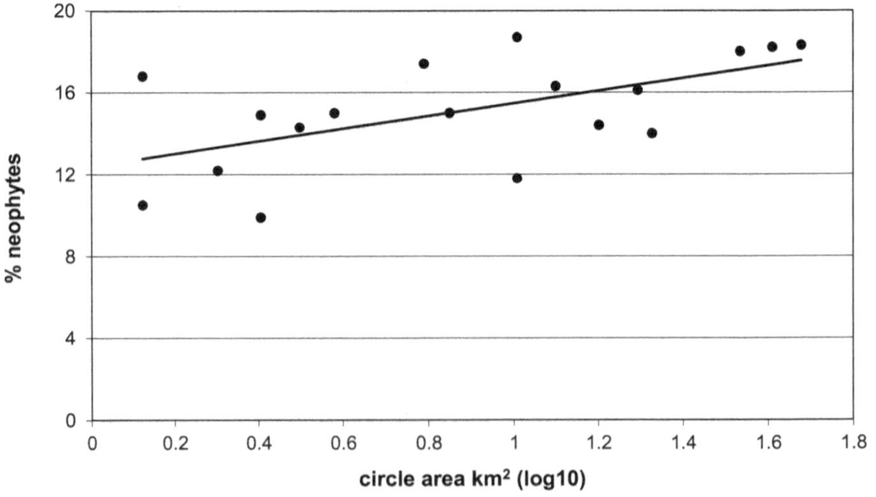

Fig. 6: The relationship of compactness and share of neophytes ($r^2 = 0.3344$, $p \leq 0.012$)

4. Conclusions

The results presented here have been obtained in the context of a test study, the aim of which was to assess in how far the overlay of data derived from municipal statistics with results from urban habitat mapping schemes allow for conclusions to be drawn on the relationship between urban form and biodiversity. The comparison of flora lists from habitat mapping schemes in a variety of cities which differ in terms of intensity of use and compactness allows for these interconnections to be studied. If there are useful interpretations to be made, these procedures can be transferred to all cities which have completed urban habitat maps.

Shrinking processes in towns are characterized by deconcentration, i.e. by the reduction of the intensity of use and by perforation as well as by a loss of spatial compactness, which is here taken to mean the dismantling of structures from the periphery inwards. By comparing cities and towns which differ in these characteristics, statements can be made on potential impacts of urban shrinking on biodiversity.

The current test study has yielded the following results:

- The number of plant species correlates positively with the size of the urban area and thus with the availability of different habitats. If the shrinking processes do not change the degree of habitat diversity, then the total number of plant species will not change either.

- A decrease in density or land-use intensity as well as an increasingly perforated urban form is likely to reduce the share of neophytes in the overall suite of species. However, neophytes represent an important share of the species richness of a city as they basically compensate for the loss of native species. If both neophytes and native species decline, it must be expected that total species numbers will also decline.

- An increase in new green spaces under the care of parks departments, *e.g.* as a result of perforation processes, may lead to a relative increase in plant species with greater nitrogen requirements. Conventional green space management can reduce the share of spontaneously occurring neophytes.

- A loss of compactness of urban form can reduce the ‘degree of urbanization’ and can thus cause a decline in the share of neophytes.

- The number of neophyte species or rather the proportion of neophytes in the full spectrum of plant species appears to be the best and simplest biological indicator for the degree of urbanization.

References

Alberti, M. (2005): The Effects of Urban Patterns on Ecosystem Functions. International Regional Science Review 28 (2): 168-192.

Falinski, J. B. (1971): Flora I roslinnosc synanthropinija wsi I miast – proba analizy porownawczej. Mater. Zakl. Fitosoc. Stos. Univ. Warsz. Warszawa-Bialowieza 27: 15-37.

Frey, H. (1999): Designing the City: Towards a More Sustainable Urban Form. London.

Jenks, M. & R. Burgess (eds.) (2000): Compact Cities: Sustainable Urban Forms for Developing Countries. London.

Jenks, M. & N. Dempsey (eds.) (2005): Future Forms and Design for Sustainable Cities. Amsterdam et al.

Jenks, M., K. Williams & E. Burton (eds.) (1996): The Compact City: A Sustainable Urban Form? London.

Kühn, I. & S. Klotz (2006): Urbanization and homogenization – Comparing the floras of urban and rural areas. Biological Conservation 127: 292-300.

McKinney, M. L. (2004): Correlated non-native species richness of birds, mammals, herptiles and plants: Scale effects of area, human populations and native plants. Biological Invasions 8 (3): 415-425.

Olden, J.D., N. L. Poff & M. L. McKinney (2006): Forecasting faunal and floral homogenization associated with human population geography in North America. Biological Conservation 127: 261-271.

Pysek, P. (1998): Alien and native species in Central European urban floras: a quantitative comparison. Journal of Biogeography 25: 155-163.

Reichholf, J. H. (2007). Stadtnatur. Munich.

Sukopp, H. & R. Wittig (1998): Stadtökologie. Ein Fachbuch für Studium und Praxis. 2. Aufl., Stuttgart.

Thomas, R., ed. (2002): Sustainable Urban Design: An Environmental Approach. London.

Williams, K., E. Burton & M. Jenks (eds.) (2000): Achieving Sustainable Urban Form. London.

Wilson, E.O. & E.O. Willis (1975): Applied Biogeography. In: Cody, M.L. & J. M. Diamond (eds.): Ecology and evolution of communities. Cambridge (Mass.), p. 522-534.

Wittig, R. (2002): Siedlungsvegetation. Stuttgart.

SHRINKING CITIES – NEW URBAN 'SOCIO-NATURES'?

Tobias Chilla

1. Changing natures

In Germany, phenomena of shrinkage are increasingly affecting former 'boom regions' characterised by economic growth and urban sprawl. After decades of predominant growth, these processes now challenge various planning paradigms and concepts. This is obviously true for issues of 'nature' in urban shrinkage contexts, too. It might be too early for a comprehensive analysis – however, it appears to be a good opportunity for a first, exploratory conceptual reflection.

Western societies historically refer to the idea of a 'dichotomy' of nature and culture. These categories appear to be antonyms; they might even exclude each other (CHILLA 2005a). The following two examples show that the idea of a dichotomy has been challenged repeatedly: Firstly, castles that historically served military purposes were solidly built. After their military function became redundant ivy covered their walls and made them look 'romantic' – and this clichéd 'visiotype' has coined our understanding of the epoch of romanticism and of its regions like the *Mittelrheintal*. Secondly, after the fall of the Iron Curtain, nature conservation has focused on the former frontier areas and started to protect them as a major 'green belt' throughout Europe (ENGELS et al. 2004).

The dichotomy of nature and culture and therein, of nature and *cities*, has traditionally been symbolized by city walls. During the last century, this dichotomy has been challenged, too: Urban sprawl blurred the border between the urban – which is mostly regarded as 'non-nature' – and the rural, which is almost a synonym for 'nature'. Another prominent example of the urban context – the greening of industrial areas – will be discussed later on.

The current intense debates on demographic change, shrinking cities, ageing societies and so on, are likely to affect the understanding of (urban) nature, too; new concepts and paradigms might be established in politics, in planning studies, geography and other related fields.

This chapter will not answer whether, or to what extent, this impression is true. Instead, the rising discussions and developments will be briefly considered on a conceptual level, linking increasing discussion about urban shrinkage with debates on urban nature(s), especially applying constructivist perspectives.

2. Socio-natures?

Concepts of 'nature' and the relationship between society and nature have changed during the last decades and centuries. Citizens and a heterogeneous variety of further actors are involved in establishing particular understandings of nature. For example, the preferences of *planning professionals* change over the course of time: Urban planning, (landscape) architecture and related professions are involved with in 'constructing' natures while being subject to differing paradigms and 'zeitgeist'. Famous examples in history are the French Garden versus the English Garden, which reveal almost contrary understandings of nature. A more recent example is the pure functionalist reference to nature in post-war urban planning in contrast to the euphoric recovery of 'ecological nature' such as green yards, roofs and facades during the 1980s.

After times of little conceptual thought about urban natures, several academics have recently focused on behavioural approaches that scrutinize *citizens'* perception of the environment (e.g. RINK 2005) and raise the question of what is perceived as nature and how it is evaluated. Others have normatively supported 'greening the city' and a new 'wilderness of the city' (such as the *Kasseler Schule*, see KÖRNER 2003).

During the last decade, much research has been conducted on the 'construction of nature' (cp. DEMERITT 2002). Following this approach, 'nature' is not simply a physical category but also a social construction: Societal, academic, and political discourses define, evaluate and construct competing types of 'nature(s)'. Particular issues are brought to public attention by the media, political forums, and scientific papers and so on.

Following this perspective, 'socio-nature' is a satisfactory term to describe the significant interweaving of society and nature (CASTREE & BRAUN 2001). This is not to neglect the material or ecologic dimension of nature but to stress the insight that the 'relevance' of nature depends to a large extent on societal discourses. This approach can be briefly introduced by explaining some crucial notions:

- As soon as a particular issue is established and the public is communicating about it in a certain way, it can be considered as a *framework*.
- These frameworks rely on certain arguments that most people believe – the notion of the '*storyline*' stresses the fact that arguments are not effective by themselves, but must be established to be believed and shared (similar: *narrative*).

- A *Masterframe* is the paradigmatic background (or 'zeitgeist'), which is a pre-condition to certain frameworks at a particular time.
- A *visiotype* is a key stylistic device that promotes certain aesthetic styles and motifs (cp. PÖRKSEN 1997).
- A *semantic field* is a typical word choice, connected with particular meanings and associations.

During recent years, a broad variety of studies has been published about biodiversity, ecologic problems in developing countries and so on. A notable strand of research has also focused on 'urban natures' (cp. CHILLA 2005b, KEIL & GRAHAM 1998), for example:

- the cultural meaning of lawns (FEAGAN & RIPMEESTER 1999)
- the interpretation of a glass of New York City water (SYNGEDOUW 1996)
- societal definitions of public parks (KÖRNER & PILGRIM 1998)
- urban and environmental history of New York City (GANDY 2002)
- New Urbanism, Sustainable Development as urbanistic models (e.g. TILL 2001)

These studies frequently reveal hidden interests of public administrations, processes of social exclusion or competing economic aims. At the same time, almost any definition and construction of urban nature *refers* to established meanings – even if it contradicts and attempts to redefine them. This is true for the academic community, too, as even highest academic expertise does not attempt to be 'neutral', but always takes part in a societal discourse.

The constructivist approach of focussing on 'socio-nature' might be, on the one hand, just another popular approach currently in vogue. On the other hand, it could be a concept that provides the opportunity to integrate the different societal actors (including politicians, media etc.) and remains *reflective* about the scientific context.

The question now is to what extent do the intensifying processes of urban shrinkage breed new constructions of 'urban nature'?

3. Urban shrinkage

The Third International Conference on Urban Ecology in Berlin, in September 2006, was one of the first public forums to address nature in the context of shrinkage (*Shrinking Cities: Ecological Consequences and Challenges for Urban Development*). It is not surprising that this conference was held in Germany: The German discourse on ecology, environment etc. has always been intense. Moreover, East

Germany has already experienced shrinking processes to a notable extent: following reunification, migration into the West and economic restructuring resulted in severe population loss.
Before addressing the consequences of shrinkage for 'nature', we should first reflect on the reasons for urban shrinkage as they are categorised within literature (e.g. KAUFMANN 2005:100 ff.).
Firstly, processes of suburbanisation do not only urbanize formerly rural areas, they also impose a more extensive structure on urban areas. Consequently these processes can result in shrinkage on a local, intraregional scale. On a wider scale, however, suburbanisation still indicates growth, so in the following discussion the focus is on further reasons for shrinkage.
Deindustrialisation is the second reason for shrinkage: In Germany, deindustrialisation first hit the Ruhr and Saar regions during the 1980s. At that time, shrinkage was predominantly considered as an economic crisis. During the 1990s this development was also recognised as a planning challenge. In some cities of these regions, deindustrialisation induced serious population losses and therefore urban shrinkage. Nevertheless, this process has rarely been labelled 'urban shrinkage' but instead as deindustrialisation or structural change.
'Demographic change' can be considered as the third reason for shrinkage. During the 1990s, shrinking cities in Eastern Germany were described as a consequence of migration losses. A few years later, shrinkage was also discussed in the context of Western cities, and this development is largely explained by a very low birth-rate.
Before returning to the latter aspect, the following section highlights the discourse on 'post-industrial natures'.

4. Post-industrial nature

Deindustrialisation – especially in the Ruhr area – was a starting point for greening industrial architecture and surrounding areas. During the 1990s the term '*Industrienatur*' was established, which is restricted to the German language. *Industrienatur* describes the growth of vegetation and reestablishment of waterways in formerly industrial areas; therefore the term 'industrial nature' is not an exact translation. This term was coined in the Ruhr area, where the re-use of manufacturing sites as leisure sites, designed with nature, was an important tool during the 1990s. The most popular examples are the Landschaftspark Duisburg Nord and Zeche Zollverein (Essen) where both planned vegetation and natural succession were important aesthetic instruments (e.g. LATZ et al. 1999). At that time, urban shrinkage and economic crises hindered the establishment of new patterns of use, and regional planning institutions accepted deindustrialisation as a 'trend of no return'.

The public sector introduced *Industrienatur*, mainly by means of major funding during the *Internationale Bauausstellung* (the International Planning Exhibition in Germany, 1989-1999). *Industrienatur* was established as a part of *Industrie-Kultur* (industrial culture), which mainly focuses on industrial architecture.
Nowdays, brochures and web sites promote *Industrienatur* as an interesting issue for both tourists and locals (see Fig. 1). The advertising materials show enhanced images and 'aesthetic upgrading' rather than photographs of abandonment and waste land. The differences between yesterday and today are often stressed with contrasting images.

Simultaneously, the notion of 'wilderness' plays an important role within the narrative of *Industrienatur*. For example, brochures present 'jungles' at places where there used to be factories, and forestry on former industrial areas is labelled as "wild urban woodlands" (Kowarik & Körner 2005).

These references aim to avoid connotations of abandonment, dilapidation or no-go-areas. Instead, a romantic, and somehow spectacular, image is promoted. Furthermore, environmental arguments play an important role.

The public focus on deindustrialisation processes has now shifted to Eastern Germany where – again – an *Internationale Bauaustellung* aims to redefine a former industrial landscape by means of *Industrienatur* (IBA Fürst Pückler Land, Lausitz). This time, former brown-coal sites are materially reshaped as leisure destinations, nature conservation areas, etc. At the same time and on a discursive level, this project promotes a post-industrial, attractive land of industrial lakes (see below Fig. 2).
In addition to these predominantly top-down planning ventures, there are examples of 'grassroots' developments. The most prominent example in Germany is Berlin Gleisdreieck, a former railway area in the middle of Berlin (www.gleisdreieck.de). In this case, private activists are calling for the area to be made suitable for use by children, immigrants, and older people. The planning aim – a 'citizens' park' – is linked to the environment, participation and urban nature. This example shows that 'shrinkage nature' is a political issue, and that there are different ways to define and promote it.

Recapitulating, we can indeed state that there is a framework of *Industrienatur* in Germany. It has been placed on the agenda mainly by the public sector, geographically focusing on the Ruhr area and now shifting to Eastern regions, and it features a coherent framework and storyline.

Fig. 1: Brochure about 'Industrienatur' (source: KVR n.d.)

5. Shrinkage natures?

As discussed earlier, processes of industrial and demographic shrinkage have occurred in Germany during recent years. Particular forms and understandings of

Fig. 2: 'Shrinkage natures?' – different scales and contexts:
upper left: vacancy in Dresden (Photography: STEFANIE RÖSSLER),
upper right: 'urban natures' between pre-fab tower blocks in Dresden (Photography: STEFANIE RÖSSLER),
lower picture: former brown-coal field near Cottbus, today part of the Project Fürst Pückler-Land (Photography: THOMAS KLÄBER / IBA)

nature have been simultaneously established. However, they are usually linked to '(post) industry' rather than 'shrinkage'. During the coming decades, many German regions will face severe shrinkage processes that are not predominantly linked to deindustrialisation processes. The question now is: how will shrinkage due to demographic change affect notions of nature? Is a new framework arising?

Literature on demographic change in Germany is currently proliferating. One of the most prominent publications on this issue is the report of the Berlin-Institut on the demographic situation in Germany (KRÖHNERT et al. 2005). This report summarises the situation with twelve statements, the last of which is titled *Wolf Succeeding Humans* ("Nach dem Menschen kommt der Wolf"). This section appears to be an anti-urban concept of the future whereby the growth of wolf and lynx populations is contrasted to the declining human population (KRÖHNERT et al. 2005:47). The report recommends using shrinking regions for "reestablishment of nature", "regeneration for nature" and for "retention areas" (ibid.). These statements are not explicitly aimed at urban contexts but they might be considered as early concepts of 'shrinkage nature'.

As 'shrinkage natures' are not established as a coherent notion of science or planning policy little work has been conducted on the field (Heiland et al. 2005; Troge & Hutter 2004; see the paper by Stefanie Rößler in this book). Therefore predicting the outcomes of the emerging discussion is difficult. Despite this uncertainty, and with regard to conceptualisation, some potential trends can already be considered.

Analysis of current literature shows that there are positive and negative assessments of shrinkage processes. These assessments can be assigned to particular disciplinary *perspectives*:
Firstly, from the perspective of architecture and planning, shrinkage can optimistically be regarded as a window of opportunity providing new scope for design or, from a more pessimistic point of view, result in fragmentation, 'muddling through', and chaos.
Secondly, from a political and administrative perspective, planning conflicts may be mitigated if economic pressure on free space decreases. On the other hand, economic crises may detract from ecological concerns (cp. Heiland et al. 2005).
Thirdly, shrinkage also influences the perspective of ecology: An important strand of nature conservation literature discusses whether nature conservation can be performed by means of more static instruments such as permanent area protection, or if more dynamic instruments should be applied. The rising debate on demographic change and urban shrinkage has become a further argument to support the dynamic approach. Urban brownfield sites with an uncertain future are especially considered to be important opportunities to develop and test more dynamic administrative planning and de facto use of urban nature (see e.g. Ecologic 2006:75 ff.). Thus, deregulation and transition to flexible urban governance are the crucial items of this discussion.
On a more general level, demographic change may result in more space for 'natural development', or the decrease in funding will result in 'monotony' due to natural succession.
Finally, the challenge of anticipating the *geographical* focus of future discourses has to be considered. It is very likely that Eastern regions will play an important role, but the Western regions are expected to become integrated into the discourse, too. Regarding the spatial context, will the main interest lie on rural or urban areas, or to put it as Heiland et al. (2005), will the interest lie in shrinking cities within prospering surroundings or prospering cities in shrinking regions etc.?

Moreover, the question of scale appears to be crucial. At least three degrees of scale can be identified (see Fig. 2): The *micro*-scale is typically of a scientific nature and focuses on individual species of plants or animals, parts of particular buildings etc. The *meso*-scale concentrates on groups of buildings: larger areas with a variety of species, a typical approach of both architects and nature scientists. Finally, the *macro*-scale is applied to whole 'shrinking' cities and regions. This scale can be considered as typical for regional planning, politics and especially for mass-media discourses.
Coherent concepts of 'shrinkage natures' have not yet been established. Which stylistic devices will be used? Will the references to 'wilderness' be established further? Will the fears of abandonment and decline be articulated? Understanding of 'shrinkage natures' might be restricted to technical issues and could be defined as simply a matter of infrastructure; the discussion may be restricted to administrative and organisational problems. A variety of possible discursive developments can be considered and expected.
The question remains of how prominently the issue of 'shrinkage natures' will be placed on the agenda, (and by whom and when). Since there is commercial interest it would not be surprising if the public sector took control of the issue. This is a question of politics, of course, but also of academic debate – so we are all involved in the process.

References

Castree, N. & B. Braun (eds.) (2001): Social nature. Malden, Oxford.

Chilla, T. (2005a): Stadt und Natur – Dichotomie, Kontinuum, soziale Konstruktion. Raumforschung und Raumordnung 63 (3): 179-188.

Chilla, T. (2005b): ‚Stadt-Naturen' in der Diskursanalyse. Konzeptionelle Hintergründe und empirische Möglichkeiten. Geographische Zeitschrift 93 (3): 183-196.

Demeritt, D. (2002): What is the ‚social construction of nature'? A typology and sympathetic critique. Progress in Human Geography 26 (6): 767-790.

Ecologic (2006): Naturschutz. Naturerbe – eine Frage der Kultur In: Hauff, V. & G. Bachmann (eds.): Unterm Strich. Erbschaften und Erblasten für das Deutschland von morgen – eine Generationenbilanz. Bonn: 75-82.

Engels, B., Heidrich, A., Nauber, J., Riecken, U., Schmauder, H. & K. Ullrich (eds.) (2004): Perspectives of the Green Belt. Chances for an ecological network from the Barents Sea to the Adriatic Sea. BfN-Skripten 102, Bonn.

Feagan, R. B. & M. Ripmeester (1999): Contesting natural(ized) lawns: a geography of private green space in the Niagara region. Urban Geography 20 (7): 617-634.

GANDY, M. (2002): Concrete and clay. Reworking nature in New York City. Cambridge, London.

HEILAND, S., REGENER, M. & S. STUTZRIEMER (2005): Auswirkungen des demographischen Wandels auf Umwelt- und Naturschutz. Raumforschung und Raumordnung 63 (3): 189-198.

KAUFMANN, F.-X. (2005): Schrumpfende Gesellschaft. Vom Bevölkerungsrückgang und seinen Folgen. Bonn.

KEIL, R. & J. GRAHAM (1998): Reasserting nature. Constructing urban environments after fordism In: Braun, B. & N. Castree (eds.): Remaking reality: nature at the millennium. London, New York: 100-125.

KÖRNER, S. (2003): Das Heimische, das Fremde, das Triviale und das Exotische: Stadtnaturbilder als Bewertungsgrundlage im Naturschutz und in der Planung. Berichte zur deutschen Landeskunde 77 (4): 349-362.

KÖRNER, W. & C. PILGRIM (1998): Diskurs und Hegemonie – Deutungsstrategien in der Frankfurter Stadtentwicklungspolitik. Geographica Helvetica 53 (3): 96-102.

KOWARIK, I. & S. KÖRNER (eds.) (2005): Wild urban woodlands. New perspectives for Urban Forestry. Berlin, Heidelberg.

KRÖHNERT, S., MEDICUS, F. & R. KLINGHOLZ (2006). Die demographische Lage der Nation. Wie zukunftsfähig sind Deutschlands Regionen? München.

KVR = Kommunalverband Ruhrgebiet (n.d.): Route Industrienatur. Brochure, Essen. Online available: http://www.route-industriekultur.de/media/downloads/pdf/RVRSammel_PDF.pdf.

LATZ, P., NEISS, T., SCHMID, A., DETTMAR, J. & K. GANSER (eds.) (1999): IndustrieNatur. Ökologie und Gartenkunst im Emscher Park. Stuttgart.

PÖRKSEN, U. (1997): Weltmarkt der Bilder. Eine Philosophie der Visiotype. Stuttgart.

RINK, D. (2005): Surrogate nature or wilderness? Social perceptions and notions of nature in an urban context In: Kowarik, I. & S. Körner (eds.): Urban wild woodlands. Berlin, Heidelberg: 67-80.

SWYNGEDOUW, E. (1996): The city as a hybrid: on nature, society and cyborg urbanization. Capitalism Nature Society 7 (2): 65-80.

TILL, K. E. (2001): New Urbansim and Nature: Green Marketing and the neotraditional community. Urban Geography 22 (3): 220-248.

TROGE, A. & C.-P. HUTTER (eds.) (2004): Der Bevölkerungsrückgang. Konsequenzen für Flächenmanagement und Umwelt. Stuttgart.

CURRENT TRENDS IN URBAN REGENERATION AND RESTRUCTURING IN GERMANY AND THEIR EFFECTS ON SUSTAINABILITY

Uwe Altrock

1. Introduction

If we agree with PETER ROBERTS' (2000:17) view of urban regeneration as a "*comprehensive and integrated vision and action which leads to the resolution of urban problems and which seeks to bring about a lasting improvement in the economic, physical, social and environmental condition of an area that has been subject to change*", it is clear that the concept includes a whole range of different strategies and measures. As a rule, these are applied to a spatially defined area (urban neighbourhood, *quartier*) and are intended to further the improvement of this area's sustainability balance through publicly funded projects. In this way a broad range of problems from land-use conflicts, problems with the condition of existing buildings, low living standards and poor working conditions due to investment backlogs through traffic and infrastructure problems to selective migration are intended to be addressed. A few decades ago it was sometimes assumed that publicly funded urban redevelopment with a high level of subsidy input would secure property values in the medium- to long-term and achieve officially specified, quantitatively measurable standards for housing and its immediate surroundings. However, experience has showed that conventional approaches contribute to the displacement of important target groups, damage their social contexts and destroy historically significant building fabric. They fail to lead disadvantaged areas out of dependence on public subsidy programmes in the medium-term and tend to weaken the local economies of urban neighbourhoods rather than strengthen them. In the face of this experience, a number of new approaches to urban regeneration with a marked change of direction have emerged since the 1990s. These approaches will be examined in more detail in the following, emphasising their effects on sustainability. Although their results are quite varied, some of the cardinal errors of older approaches such as the implied fixation on the construction industry could be moderated or removed. The planning instruments have thereby constantly adapted to the changing political perception of the problems to be dealt with.

It will become clear that in an overall political climate of increasing individualisation and relatively cautious regulation of the locational choices of households and companies by the market economy, planning policies cannot reverse an overall societal trend that favours overdevelopment and urban sprawl. Nevertheless, it is possible to encourage certain more recent trends and thereby to preserve and develop some of the strengths of the compact 'European city' under difficult conditions. A resigned attitude, unquestioningly accepting the supposed dissolution of our cities, seems incomprehensible on the basis of these findings. Nevertheless, it will become apparent that there is only indirect potential for positive contributions from urban regeneration policy to ecological sustainability and these rely on persistent communication of these indirect effects to the general public in order to prevent them being thoughtlessly sacrificed to supposedly simpler political concepts.

2. The established approaches and their perspectives: Careful Urban Regeneration, the preservation of historic neighbourhoods, conversion

Careful Urban Regeneration in old inner city neighbourhoods and its application in East Germany

With heavy societal criticism of urban renewal and its massive demolition of building stock in the late modern period, the paradigm of so-called 'Careful Urban Regeneration' was born in numerous large cities in staggered phases and sometimes mutually inspired in the late 1970s and early 1980s. This paradigm continues to dominate today, especially in Central Europe. It is characterised by the concentration of funding on the renovation and modernisation of housing with a strong orientation towards existing building stock, measures to improve public spaces, the maintenance of cheaper housing, a strong tenant-orientation and a declaration of belief in the preservation of social contexts and thus the least possible displacement of the traditional population.

A significant achievement of Careful Urban Regeneration is the fact that urban policy since German unification has addressed the repair of long-neglected older urban neighbourhoods as well as (initially) the improvement of newer housing complexes, with a strong emphasis on maintaining existing buildings. As in the inner city neighbourhoods in West Germany, there has been considerable success in East Germany in making structurally intact neighbourhoods liveable again with the aid of large subsidies, making them attractive to large sections of the population. In the old Federal Republic it had been necessary to implement this orientation towards existing building stock against powerful alliances in politics and often also in public housing companies in order to prevent a 'second destruction' of German inner cities after the bombing of the Second World War (Bodenschatz et al. 1983, Autzen 1984, Bodenschatz 1987). In the East German *Länder* the main issue was how the backlog of maintenance and modernisation work built up over decades could be worked through in a short period of time. In the meantime it had been demonstrated that the use of public subsidies in urban regeneration is the best-suited form of public funding to stimulate private investment (Lessat 1996, Blume 1999) and the principle of urban development subsidies was not yet being politically questioned by the early, still tentative neo-liberalisation wave of the 1980s. In West Germany it had however, also become apparent that the process of urban regeneration in the neighbourhoods affected was much more costly than initially believed (Autzen et al. 1986). Thus dealing with the exceptional problems in East Germany was still an unresolved issue.

The dramatically altered emphasis in urban development funding and the now established orientation towards the maintenance of existing buildings implied the application of the proven subsidising instruments, allowing comprehensive securing and improvement of the existing built fabric. Indeed, the outstanding quality of the existing urban substance allowed for the introduction of more extensive programmes (see below on the preservation of built urban heritage). Nevertheless, the general conditions for urban regeneration changed drastically, at least in comparison with the elaborate model programmes of the 1980s such as the International Building Exhibition (IBA) in Berlin in 1987 (Bernt 2003). The very large extent of the areas to be improved meant that urban regeneration had to do without expensive but tried and tested measures for the redevelopment of public spaces and in some cases also had to limit measures for the upgrading of the built fabric. Local governments had to accept their diminishing influence on the details of the urban regeneration process which was increasingly shaped by private investors and was counteracted by an increasingly landlord-friendly housing policy, the extensive abolition of social housing and far-reaching legal limitations on the use of upper limits for rents. The stabilisation of old inner city areas was impeded by other problems as well and took place slowly.

The introduction of the preservation of historic neighbourhoods in East Germany

The large number of intact historical urban centres in East Germany (although these had a badly neglected appearance at the time of German unification) led to the introduction of a special programme to secure the built fabric and renew urban ensembles in a manner consistent with the preservation of historic buildings. This programme succeeded in stabilising and upgrading a large number of centres of small towns in particular in an exemplary manner (www. staedtebaulicher-denkmalschutz.de, BM-Bau various issues, BMVBW & IRS 2004, Kuder 2007). A number of procedural innovations were introduced by the programme – such as the accompaniment of the regeneration process by a group of experts and regular conferences for reporting on achievements and to facilitate the exchange of experiences. Although the programme was widely perceived as a success, in many cases it contributed to the preservation of town and city centres but could not solve the fundamental problems of East German inner cities and town centres. These continue to be characterised by a lack of vitality, partly due to the difficulty of adapting historic buildings to modern uses, and the weakness of medium-sized retailing businesses in East Germany. Gradually, however, the positive effects emanating from the renewed urban centres are also contributing to a renaissance of the smaller town centres, thanks to an increase in visitor numbers and moderate growth in the numbers of town centre residents. In spite of the many successes, both careful urban regeneration and the protection of urban heritage have not been able to reduce the calls for more socially and economically oriented urban regeneration.

The internal development, the trend towards brownfield development and the restructuring of conversion sites

As early as the late 1980s, and later to a greater extent in the course of accelerated structural transformation and the end of the Cold War, large sites in towns and cities went out of use which had for decades been exclusively reserved as more or less enclosed enclaves for specific uses. When these military, infrastructural and industrial sites were relinquished, the conversion of these sites made it possible to dispense with zoning greenfield sites for new construction. In this way it was often possible to regain interesting key locations in the towns and cities, such as waterfront locations, for public use. New uses of historically significant buildings gave the newly emerging neighbourhoods a particular, individual character with positive effects for their marketability. The availability of large municipally owned sites also made it possible to re-use these inexpensively as public open spaces without having to demolish existing buildings in other locations.

Consequently, the conversion of such sites became a key instrument in the success of internal urban development, the effects of which can hardly be overestimated. It should, however, be borne in mind that the re-use of existing buildings and sites sometimes presents developers with major – and also financial – challenges not always easily met, and on occasion leads to conflicts, e.g. with regard to whether old industrial buildings were worthy of maintenance or preservation. Some conversion projects could only be realised after painfully slow efforts at persuasion as the new users would have preferred to move into new buildings instead of into military barracks or industrial buildings that had long had a negative image. Economic transformation, which was an important driving force making these brownfield sites available in the first place, fortunately also had a positive side insofar as many newly founded service industries and creative businesses particularly valued the historical ambience of the old buildings. As a result neighbourhoods developed that meet important planning requirements for sustainability in an exemplary fashion. These frequently award-winning projects cannot all be listed here, so great is their number. It is now apparent that some of the inner city locations available for conversion have become sure-fire successes and particularly attract media and creative businesses, while some peripheral locations – especially former military properties – are seriously limited in the range of new uses applicable because of their unfavourable locations.

Brownfield site conversion has been supported mainly by the *Länder* most affected. Favourable locations could and can be converted without intensive subsidisation. Where potential and demand were overestimated, as in the case of the Urban Development Measures in Berlin in the 1990s, locations stagnate for years and swallow large quantities of public funding – especially when they become a comfortable location to 'park' less dynamic staff or are used to implement long obsolete urban visions (Altrock 2000).

Perspectives/development potential

A review of the three established instruments for urban regeneration shows that in the past a large number of urban neighbourhoods has already been preserved, upgraded and converted. However, experts' attention is now directed towards other fields of action which will be presented in greater detail in the next section. This is partly due to the direction of funding, perceived pressure to take action and uncertainty in addressing new challenges. Careful Urban Regeneration in individual neighbourhoods, the preservation of urban heritage and conversion will all continue to play a significant role in the development of existing urban fabric. The concept of carefulness is still undisputed, although in some urban neighbourhoods with high vacancy levels planners are reconsidering the application of more radical measures. Conversion has not outlived its usefulness, especially as new areas are still affected by structural change and fall into disuse. It is increasingly being included in urban redevelopment and is reaching formerly vigorous small and medium-sized towns and cities (see the *Stadtumbau West* ('Urban Restructuring West') programme in particular).

3. Current challenges: Neighbourhood management and urban redevelopment

Endeavours to integrate social policies with urban neighbourhood development and the introduction of neighbourhood management

From the early 1990s it was increasingly recognised that conventional, construction-oriented urban regeneration approaches did not pay enough attention to the socio-economic problems of disadvantaged neighbourhoods and it was concluded that the orientation of such planning must be changed in favour of greater consideration of social networks, the local economy and training the local population. Such integrative policy approaches had already been tested in an embryonic form in some towns and cities in the framework of the Careful Urban Regeneration of the 1980s. However they only made their breakthrough in the late 1990s, inspired by experience with the European URBAN programme, and were tested for the first time in North Rhine-Westphalia, Hamburg and Berlin. The introduction of this initiative at a national level, familiar under the name 'Socially Integrative City', took this changed orientation into account. Using a relatively modest level of funding per neighbourhood this initiative applied upgrading measures in an attempt to counteract social polarisation and the selective out-migration of higher income residents (Häussermann & Institut für Stadtforschung und Strukturpolitik et al. 1998). The introduction of neighbourhood management, aimed at consolidation of the new approaches with intensive citizen participation, managed to close a "gap in the chart of subsidy policies".

The aim of the programme to halt a downward social spiral or prevent it developing in the first place could only be addressed very indirectly. The instruments introduced for this purpose, which emphasise an intensive, citizen-oriented participation strategy for the development and implementation of an integrated action plan with reference to neighbourhood-oriented social and economic policy, can indeed help to improve living conditions locally but do very little to change structural problems which have their roots beyond the neighbourhoods, e.g. in the effects of general economic structural change. Nevertheless the innovations in procedures and measures that were catalysed by the programme are considerable. Among the many contributions job training measures for problem groups, youths and migrants (especially women) should be mentioned, as well as platforms to let commercial spaces, and commercial street management including innovative retail projects in the artistic sector, support for intermediate uses which was still quite rare in the late 1990s, or support for local initiatives already in existence, e.g. in applying for subsidies.

There was also a series of procedural innovations with varying degrees of influence. Conflict resolution procedures addressed central public spaces and squares in neighbourhoods, which combine very different uses that are difficult to reconcile with one another (e.g. families with children, dog owners, homeless people, preservation of historic buildings and monuments). The planning procedure used here differs from earlier ones in that it no longer attempts to implement fixed quantitative guidelines as a measure of quality – which in the example of conflicts between the users of open spaces could have led to an increase in the number of open spaces through demolition measures in high density inner city neighbourhoods –, but instead organises the immediate neighbourhood of the various users in the form of shared use to reduce conflicts or breaks up different types of use at a moderate temporal or spatial level.

The most significant procedural innovation which takes changed societal conditions into account is the introduction of neighbourhood funds. Here local government allows the residents of the neighbourhood to decide how part of the available subsidies will be spent. Here participation does not only involve consultation and the incorporation of suggestions, but real inclusion in programmatic orientation. The handing over of responsibility to the neighbourhood residents is organised by neighbourhood management and is associated with an attempt at a civil society reorientation of urban policy. The responsible distribution of the available funds by the selected citizens in a committee established especially for this purpose affirms this approach, but the funds available at this level are very small in comparison with municipal investments, for example. Ultimately this is a group of interesting subsidised small projects, which although they are of central significance for the neighbourhood still cannot solve the societally determined social problems and impediments to economic development that culminate in the respective neighbourhood.

To sum up the effects of the Socially Integrative City programme and other similar approaches (Aehnelt 2005, Walther 2002, Austermann & Zimmer-Hegmann 2000, Häussermann 2005), the contribution of the neighbourhood management approach to the solution of social crises is quite modest. While this doesn't speak against it, it does question the frequently high expectations of this approach. Only in the long-term can it make a fundamental contribution to the demographic and economic stabilisation of a neighbourhood. The achievements of neighbourhood management can also be measured in the number of problematic but nevertheless liveable neighbourhoods that are by no means irretrievable which are not reported on in polemical "ghetto articles" in the *Spiegel* (left-wing news weekly). Such articles not only perpetuate the stigmatisation of such neighbourhoods but can also reduce private investment to zero for months or years at one blow, thereby creating the very problems facilely described.

In spite of the positive achievements of the shift in the emphasis of funding towards social and local economy issues, neighbourhood management has not succeeded in fulfilling the hope of fundamental procedural innovation in public administration. The hoped for integration of the architectural, social policy and economic subsidy policy competences of local government met with only limited success (Güntner 2007). Neighbourhood management as a local resource for the neighbourhood residents and a transmitter of local needs to departmentally bound administration has not significantly changed conventional administrative procedures with all their problems. It is very important to see which locally developed structures and which civil society initiatives will last after the closure of neighbourhood offices when the subsidies cease.

Demographic and economic-structural changes and the *Stadtumbau Ost* ('Urban Restructuring East') and *Stadtumbau West* ('Urban Restructuring West') programmes

The dramatic changes associated with the demise of former leading industrial sectors and the decline of population in many West and East German towns and cities since the 1990s in particular are now affecting growing numbers of regions and no longer only impinge on formerly typical 'crisis regions' such as the Ruhr area. Today, even some of the small and medium-sized towns that were very prosperous in the 1980s are strongly affected and are seeking a reorientation of their urban development policy. This is all the more important as they cannot place their hopes in the usual sectoral structural change that led to an increase in the numbers employed in the service sector in large cities in particular and which would at least cushion the impact of these changes.

Because of their threat to the economic viability of entire region and sectors, the extent of the changes described brought urban development programmes into the arena and gave birth to the *Stadtumbau Ost* and *Stadtumbau West* programmes. The local consequences such as high levels of housing vacancies, underutilised infrastructure and precarious municipal budgetary situations cannot be overcome without public funding – and it is not even certain whether they can be overcome with the help of such funding. While a more detailed examination of the *Stadtumbau West* programme would be interesting, it does not have a place here as the programme is relatively limited in its effects and the approaches used do not differ so significantly from the tried and tested methods described above.

In contrast, the *Stadtumbau Ost* programme involves a paradigm shift in urban policy whose consequences can scarcely be overestimated (BMVBW & BBR 2003, Altrock 2005). It has its roots in upgrading programmes for large-scale housing projects in East Germany in the 1990s, which in turn follow on from the conservative regeneration programmes for large-scale housing projects in West Germany in the 1980s. After German unification, the then current taboo on demolition was quickly transferred to East German town planning policy by an alliance of West and East German planners. This taboo initially drew support from arguments such as the mixed social structure of East German large-scale housing projects, the high level of acceptance of these projects by their residents, the young age of the buildings and the historical conditions of the pre-unification period which had only prevented the completion of the surroundings in the neighbourhoods which were basically well-equipped with infrastructure. While at first it was possible to direct large sums of money into the upgrading of the prefabricated housing projects, by the late 1990s the consequences of a dual out-migration trend into suburban single-family houses and to the West German agglomerations with their employment opportunities were felt in high vacancy levels. The continuing high level of acceptance of these housing complexes did not help to prevent this migration and eventually its growing extent became an existential threat to some public and semi-public housing companies.

In the Federal-*Länder Stadtumbau Ost* programme, it becomes apparent for perhaps the first time how the political field of urban development policy, which developed in its present form in the context of the need to deal with urban growth, can be restructured in an experimental way and with the involvement of all the relevant groups of actors. Although the earlier approaches to urban regeneration had dealt with small-scale decline it had always seemed possible to aim for small-scale revitalisation using public resources and the private investment thus stimulated. This was an important aim of urban regeneration policy together with the improvement of living conditions for the local population. The intention was to stabilise property values for local owners, which they could not manage individually in periods of decline, making urban regeneration measures a 'public good' in the economic sense.

Integrated Urban Development Concepts and local development concepts based on these for individual districts and neighbourhoods are to be used to secure the viability of the housing market and to avoid large-scale dereliction as a result of the housing demolition necessary because of high vacancy levels, ensuring that this demolition takes places in an ordered manner and leaves behind a viable urban structure. This takes place in consultation between the housing companies, the municipal authority and other parties involved such as the companies responsible for public works. The process is supported by public funds. The intention of the programme is that the urban restructuring process has to take the overall urban perspective into account and include upgrading measures in the affected neighbourhoods in its plans. The large housing complexes on the urban peripheries are not the only ones affected, although they are suitable to a certain degree for urban reduction because of their relation to the overall urban structure and the nature of the infrastructural networks. However, buildings from other periods still remain empty or have poor long-term utilisation prospects, e.g. the inner city areas and late nineteenth century districts explicitly treated in a preferential way in the programme. Conflicts over the reduction scenario are inevitable and often unavoidable, as the interests of the housing companies differ markedly from each other and from those of the municipalities.

Fig. 1: Temporary art installation in Aschersleben by Christopher Winter, erected in the framework of urban redevelopment (part of the "drive-thru-gallery" project by chezweitz&roseapple, www.szenografie.org)

A degree of disillusionment can be observed in many places in the face of the now tatty appearance of urban peripheries, where widespread demolition has taken place on the basis of compromises between local actors. Nevertheless, as well as some spectacular refurbishments of prefabricated housing and the stabilisation of some housing companies and co-operatives, there are some interesting developments in attempts at upgrading, whose significance depends on the priorities of urban regeneration programmes and which have transformed the practice of urban development subsidisation to an extent unimaginable only a few years ago. These include the wide distribution of intermediate uses in available sites and the attempt to 'play' artistically with urban conversion sites (Fig.1). Intermediate uses already played a minor role in some earlier urban regeneration programmes. Now, however, the large number of formerly occupied sites becoming available is the reason for the toleration or even encouragement of temporary uses even in smaller towns and by conventional owners. Although these uses do not promise high returns, at least they prevent further decay or may suggest a previously unforeseeable unconventional use for the property. Performance approaches in urban redevelopment also move in a similar direction, aiming to point to creative potential subsequent uses, to preserve traces of earlier phases of use or to reinterpret them, to win over the local population for participation in the urban redevelopment process or to test completely different ways of dealing with brownfield sites for a specific period of time (ALTROCK et al. 2006). Such an approach, although it may achieve little in the face of the dramatic consequences of shrinkage in some neighbourhoods, does offer a perspective on possibilities for civil society participation and appropriation not otherwise explored. They should not become a fig leaf to hide otherwise unfair consequences of urban redevelopment processes nor should their effects be overestimated. Given the open-endedness of urban development processes during a phase of far-reaching restructuring of both the economic and physical structures of shrinking cities, they provide inspiration for and examples of participation and urban living where the usual forms of social exchange on their own could be misunderstood as a 'tranquillizer' for the remaining local residents.

4. The effects of urban regeneration and urban redevelopment on sustainability

The sustainability debate has accompanied the aspects of urban regeneration discussed here since the late 1980s. The intensive way in which urban development policy has addressed the concept of sustainability has now produced a degree of fatigue in the face of the difficulties of its operationalisation, but it is now standard. It has become clear that in spite of much progress in measuring sustainability, e.g. by means of complex systems of indicators, the main success of the sustainability debate on urban development policy and thus on urban regeneration lies in the sensitisation of the relevant actors to simultaneously taking into account economic, social and ecological aims in the process. An account of this influence on urban regeneration will conclude this article.

Firstly it should be observed that in the course of the development of the existing urban fabric, the three aspects of sustainability rarely present a full range of relevant points that strategy development can take up. The economic aspect of sustainability is taken into account in two respects. Firstly, it plays a role in the question of the financial viability of urban regeneration measures for public budgets, e.g. when examining the efficiency of investments and subventions. Studies of the stimulating effects of public urban development funding largely deprive the issue of its drama. Only in connection with the preservation of historic buildings and monuments and urban reduction in the framework of *Stadtumbau Ost* is this issue being currently seriously discussed. Secondly, especially after the increased attention paid to urban neighbourhoods in crisis in the early 1990s (FROESSLER 1994) and the associated spotlight on neighbourhood-specific employment issues, urban development initiatives increasingly attempted to push local economies. An end of this development is not yet in sight, in spite of often only modest successes. The central problem in the encouragement of local economies lies in the fact that neither workers nor employers are organised on a neighbourhood basis. Opportunities for small-scale economic subventions include developing human capital through training measures, finding commercial sites in the neighbourhood and creating favourable conditions for business start-ups. Even in disadvantaged areas where the residents' attachment to their neighbourhood is relatively strong, intensive employment interconnections exist with other neighbourhoods. Small-scale projects emphasising hard-to-place long-term unemployed citizens are a very important supplement to general urban economic stimulation and some observations show that they can indeed have a significant effect on the employment situation (Kaufmann 2007). The main relevant effects on the number of jobs and economic influences are however a product of exogenous events in the city as a whole or nationwide and can scarcely be influenced by urban development policies. In particular it is apparent that the economic disadvantages of urban neighbourhoods can only be changed politically in the long-term or not at all. Especially since the 'discovery' of Careful Urban Regeneration the living conditions of local residents and their social context have been at the centre of the urban regeneration policy debate. Academic criticism of the results of urban policy has repeatedly been directed against its effects on social justice and thus its influence on the creation and exacerbation of and its insufficient concern with social polarisation in the city. Urban policy, like other areas of policy, must always deal with the tension between economic and social effects. While the establishment of a meaningful balance between growth policy and distribution policy is always a central issue, it must also be recognised that the creation of growth impulses in other parts of the city is not usually an adequate instrument to improve the situation of the disadvantaged. If we look at the effects of Careful Urban Regeneration and neighbourhood management, there are some indications that they can make a useful contribution to the creation of the necessary balance in urban policy, if CBD, scientific and site policies for the city centre and good locations for innovative businesses are usefully combined with neighbourhood policy using the instruments described above. In the case of urban redevelopment, which can only make limited use of the existing potential for local economic promotion in the face of decreasing funding, it is more difficult to assess the situation.

In the case of ecology, urban regeneration only makes very small-scale concessions. They range from the proven model projects for the energy-saving and resource-efficient redevelopment of high-density late-nineteenth century blocks practised since the 1980s in the framework of the International Building Exhibition (IBA) in Berlin, to the improvement of neighbourhood green spaces. Although in individual cases contributions can be made to ecological improvement and this has now become standard, the overall effects are mostly indirect. Urban regeneration is ecological in that it keeps neighbourhoods attractive and prevents out-migration to the suburbs. 'Green' achievements such as the creation of neighbourhood parks on former railway sites do bring direct ecological gain, but this is probably negligible in comparison with the improvement of the quality of life for various sectors of the population. If neighbourhoods are so improved that the population density falls as a result of changes in the residential structure (rising incomes and higher consumption of living space per inhabitant) the upgrading measures can even have negative ecological side effects.

If we look at the reasons for selective out-migration from inner city neighbourhoods it is clear that these have serious difficulties in getting rid of their manifest disadvantages for certain population groups. However, this is a precondition for their continued contribution to the avoidance of overdevelopment and urban sprawl as described above, and thereby for their qualification as ecologically sustainable. There are now at least some attempts to provide an attractive quality of living in the city for families or residents who value private open space comparable to that on the urban periphery. Schools policy, which should contribute to preventing families with children of school-going age from moving to 'better' school districts, has not yet made that much progress. Here it is apparent that only a complex policy that can maintain the attractiveness of urban districts for broad sections of the population without excluding disadvantaged groups in combination with economic, social and economic requirements and approaches really deserves to be called sustainable – but it is also clear that it is very difficult to implement such a policy at a time characterised by a dwindling scope of action for local government.

It is still a few outstanding conversion projects of the 1990s that can continue to claim exemplary standards, e.g. the conversion of the French Quarter in Tübingen (Feldtkeller 2001). Under favourable circumstances these projects can provide high densities and an intensive mixture of uses with a wide range of employment opportunities and social institutions, while at the same time creating excellent open spaces, finding new uses for existing buildings, reducing traffic and securing a degree of social variety (Fig. 2). However, they only come about when the general conditions are favourable. Of particular significance is the committed and innovative directing influence of urban regeneration administration, an open-minded political climate and a population willing to live in urban densities outside the large cities, which accepts the new qualities of such districts without mourning detached single-family housing. The particular contribution of urban regeneration policy lies in finding new uses as quickly as possible, in order dispense with new construction sites in greenfield locations, at least to some extent. The projects of the 1990s cited here demonstrated this. Whether they are repeatable in times of shrinkage will depend

Fig. 2: Lorettoplatz in the conversion district of Tübingen-Südstadt. Source: http://www.tuebingen.de/bilddatenbank/Lorettoplatz_v_Osten2.jpg, accessed 19.06.2007

on which arguments win out in the affected towns and cities: the plea for a radical reduction in area in the face of the stagnating or falling household numbers or for reduced expectations with regard to planning standards in new housing developments given the low level of demand.

Thus it is apparent that the redevelopment measures currently necessary in many towns and cities have left many questions open with regard to the policies for dealing with existing urban buildings and their sustainability. The positive experiences with Careful Urban Regeneration and neighbourhood management in spite of the more recent consequences of social polarisation cannot be projected into the future. The exemplary conversion projects of earlier decades are not necessarily replicable. Urban redevelopment in East and West Germany brings completely new challenges, which in the long-term could considerably limit the consumption of resources in urban life, but which also contain many traps in the course of attempts at reduction and in the worst case could lead to an un-ecological, low-density, sprawling urban structure which is not even socially or economically sustainable. The experience of the past has also shown, however, that the use of public funding in urban regeneration opens up a number of perspectives that should continue to ensure that the maintenance of existing buildings and structures will continue to play a central role in urban development policy.

References

AEHNELT, R. (2005): Zwischenevaluierung des Bund-Länder-Programms ‚Soziale Stadt' - zentrale Ergebnisse. In: Informationen zur Raumentwicklung Vol. 2/3.2005: 63-73.

ALTROCK, U. (2005): Stadtumbau in schrumpfenden Städten – Anzeichen für ein neues Governance-Modell? In: ALTROCK, U., KUNZE, R., SCHUBERT, D. & U. VON PETZ (eds.): Jahrbuch Stadterneuerung 2004/5, Berlin: 155-176.

ALTROCK, U. (2000): Die Berliner Entwicklungsgebiete - eine Kritik aus planerischer Sicht. In: ALTROCK, U. (ed.): Das Schöne im Notwendigen finden. Berlin: 183-195.

ALTROCK, U., HUNING, S., & D. PETERS (2006): Neue Wege in der Planungspraxis und warum aktuelle Planungstheorien unvollständig bleiben. In: SELLE, K. (ed.): Zur räumlichen Entwicklung beitragen. Dortmund: 248-262.

AUSTERMANN, K. & R. ZIMMER-HEGMANN (2000): Analyse der Umsetzung des integrierten Handlungsprogramms für Stadtteile mit besonderem Erneuerungsbedarf. Dortmund: ILS-Schriftenreihe No. 166.

AUTZEN, R. et al. [Bearb.] (1986): Erfahrungen mit der Sanierung nach dem Städtebauförderungsgesetz - Perspektiven der Stadterneuerung. Schriftenreihe des Bundesministers für Raumordnung, Bauwesen und Städtebau 02.036.

AUTZEN, R., BECKER, H., BODENSCHATZ, H., CLAUSSEN, H., RADICKE, D., STIMMANN, H. & M. TAEGER (1984): Stadterneuerung in Berlin. Sanierung und Zerstörung vor und neben der IBA. Berlin.

BERNT, M. (2003): Rübergeklappt. Berlin.

BLUME, L., GAULKE, K.-P. & J. ROTHER (1999): Ökonomische Wirkungen der Städtebauförderung in Mecklenburg-Vorpommern. Berlin: DIW-Beiträge zur Strukturforschung SF 182.

BODENSCHATZ, H. (1987): Platz frei für das neue Berlin!: Geschichte der Stadterneuerung in der ‚größten Mietskasernenstadt der Welt' seit 1871. Berlin.

BODENSCHATZ, H., HEISE, V. et al. (1983): Schluß mit der Zerstörung?: Stadterneuerung und städtische Opposition in West-Berlin, Amsterdam und London. Gießen.

BUNDESMINISTERIUM FÜR BAUWESEN, RAUMORDNUNG UND STÄDTEBAU (BMBau) (ed.) (verschiedene Jahrgänge): Informationsdienste städtebaulicher Denkmalschutz. Bonn.

BUNDESMINISTERIUM FÜR VERKEHR, BAU- UND WOHNUNGSWESEN (BMVBW) & Institut für Regionalentwicklung und Strukturplanung (IRS), (eds.) (2004): Ergebnisse der Begleitforschung zum Programm Städtebaulicher Denkmalschutz. Stadtumbau 2003 in den Programmstädten Städtebaulicher Denkmalschutz. Berlin.

BUNDESMINISTERIUM FÜR VERKEHR, BAU- UND WOHNUNGSWESEN (BMVBW); BUNDESAMT FÜR BAUWESEN UND RAUMORDNUNG (BBR), (eds.) (2003): Auswertung des Bundeswettbewerbs ‚Stadtumbau Ost' – für lebenswerte Städte und attraktives Wohnen. Berlin.

FELDTKELLER, A., ed. (2001): Städtebau: Vielfalt und Integration. Stuttgart/München.

FROESSLER, R. (1994): Stadtviertel in der Krise. Dortmund: ILS-Schriften, Vol. 87.

GÜNTNER, S. (2007): Soziale Stadtpolitik. Bielefeld.

HÄUSSERMANN, H. (2005): Das Programm ‚Stadtteile mit besonderem Entwicklungsbedarf - die Soziale Stadt', Gesamtbewertung und Empfehlungen der Zwischenevaluation 2003/2004. In: Informationen zur Raumentwicklung Vol. 2/3.2005: 75-85.

HÄUSSERMANN, H. & Institut für Stadtforschung und Strukturpolitik et al. (1998): Sozialorientierte Stadtentwicklung. Gutachten im Auftrag der Senatsverwaltung für Stadtentwicklung, Umweltschutz und Technologie. Berlin.

KAUFMANN, A. (2007): Paper on urban reconstruction in eastern Leipzig in the Neighbourhood Office 21.05.20007

KUDER, T. (2007): Der Städtebauliche Denkmalschutz im Wandel. – Will be published in: ALTROCK, U., KUNZE, R., PAHL-WEBER, E., SCHUBERT, D. & U. VON PETZ (Eds.): Jahrbuch Stadterneuerung 2008. Berlin.

LESSAT, V. (1996): Expertise über die Anstoßwirkungen öffentlicher Mittel in der Städtebauförderung. Berlin. DIW-Beiträge zur Strukturforschung SF 164

ROBERTS, P (2000): The Evolution, Definition and Purpose of Urban Regeneration". In: Roberts, P. & H. Sykes (eds.): Urban regeneration: a handbook. London / Thousand Oaks: 9-36.

WALTHER, U.-J., (ed.) (2002): Soziale Stadt - Zwischenbilanzen: ein Programm auf dem Weg zur sozialen Stadt? Opladen.

SIX YEARS OF *STADTUMBAU OST* ('URBAN RESTRUCTURING EAST') PROGRAMME: DIFFICULTIES OF DEALING WITH SHRINKING CITIES

Matthias Bernt

1. Introduction

In Eastern Germany ca. 1.3 million housing units are vacant. In order to deal with this problem, the German Federal Government and its *Länder* initiated the *Stadtumbau Ost* ('Urban Restructuring East') programme "for cities worth living in and attractive housing" five years ago at a cost of 2.7 billion Euros. In the framework of this programme, ca. 350,000 surplus housing units in 340 East German towns and cities are to be demolished by 2010. This demolition programme has now begun almost everywhere in East Germany and is not limited to the removal of a few eyesores – in many towns and cities whole neighbourhoods are up for consideration.

The approach adopted by the *Stadtumbau Ost* programme is new in many respects: Firstly it is no longer merely a 'catching up programme' intended to close the gap between East and West German *Länder*, but rather it reacts to a combination of problems that only developed in this particular form through the mistakes of German transformation policy in the 1990s. Secondly, for the first time in the history of German housing it calls for the demolition of housing with no provision of replacements and thus departs from the growth logic of promotion policies, at least for a limited sector. Thirdly it attempts to associate this very procedure with a target orientation towards 'sustainable urban development' (BMVBW 2005).

Thus urban restructuring in East German towns and cities can be seen as a fundamental change of direction in urban development policy, from managing growth to structuring shrinkage. First experience of structuring shrinkage is currently being gained in East Germany. Urban restructuring not only sets the course for the future of East German cities, but is likely to find imitators in a number of West German and European cities given the expected future demographic pattern of accelerating population decline (KAUFMANN 2005; see FUHRICH & KALTENBRUNNER 2005 on the West German counterpart *Stadtumbau West*).

What practical experience is being gained in *Stadtumbau Ost*? Will the problems of the East German housing market be solved? Will it be possible to implement 'sustainable urban development' after the frustrating experiences of the 1990s? Will the East indeed prove to be the avant garde (as hoped by WOLFGANG ENGLER for example, ENGLER 2002) in urban restructuring, and be the first to follow the path of urban development without growth?

This article attempts to take stock provisionally after six years of urban restructuring. To this end the East German urban crisis to which the programme is reacting will be outlined. In a second step the ambivalence and contradictions of the programme will be discussed. Thirdly, sobering conclusions will be outlined.

2. The road to crisis

If we look back ten years we can se that the leading issues of urban development in East Germany have shifted 180 degrees. In the mid-1990s the rapid improvement and expansion of housing, in the form of the renovation of decaying historic town centres and the erection of new housing, was among the top priorities of municipal policy. Supported by tax relief and state subsidies, and in reaction to the housing shortage of the time, large amounts of new housing were erected. Thus from 1991 to 1999 a total of 773,368 new housing units was created (Kommission 2000), 85 % of these being new buildings, mostly on greenfield sites on the urban peripheries.
This boom was in total opposition to the basic trends in the East German economy, as the East German economy collapsed after the German monetary union of 1990; the result of the economic decline was a wave of migration to the prosperous regions of West Germany. Between 1989 and 1999, East Germany lost ca. 1.2 million inhabitants (ca. 8 % of the population) through migration to West Germany alone. The effects of this out-migration were further exacerbated by unusual demographic features: As the numbers of births in the East German *Länder* fell to an extent unique in history, the population declined by a further 700,000. Thus on the one hand almost 0.8 million new housing units were constructed while on the other hand the number of people who could occupy these units fell by almost two million. It is obvious that such a combination can easily cause problems.
Accordingly, the construction boom was followed from 1997/98 by disillusionment. The first to feel the coming shrinkage were the housing companies. Here the gap between the growing amount of housing available and the drop in demand led to rapid growth in the numbers of vacant housing units from the mid-1990s on. This resulted in an exorbitant rise in costs and a massive decline in profits. Following intensive lobbying on the part of the housing companies, in 1998 the Federal Government finally appointed a committee of experts to analyse the problem in greater detail. Its findings, published in 2000, showed that roughly one million housing units, or about a sixth of the entire housing stock, were unoccupied in East Germany (Kommission 2000). This shocked the planning specialists and finally placed the issue on the Federal policy agenda.

3. A contradictory programme

As a result, in autumn 2001 the Federal Government established a programme that subsidises the complete demolition of housing with no replacement for the first time in the history of German housing policy: the programme *Stadtumbau Ost* (BMVBW/BBR 2003). In this programme urban development policy attempts a balancing act: On the one hand urban restructuring is clearly conceived as a consolidation programme for the local housing market with aim of rapidly eliminating surplus housing. On the other hand the programme contains ambitious urban planning goals which are supposed to ensure that the planned demolition projects are integrated in the urban structure and upgrade rather than downgrade the affected neighbourhoods. Thus the programme was divided into two sectors from the very beginning: In order to get rid of surplus housing it subsidised demolition; but this was to be linked with a simultaneous upgrading of the remaining housing, the neighbourhoods as a whole and the infrastructure. "More urbanity through fewer houses" is a slogan that describes this target in a condensed way.

Of particular significance in *Stadtumbau Ost* is the involvement of property owners. It would be impossible to implement the programme without their consent to the demolition of their property, which is protected by law. The inclusion of the property owners is thus the non plus ultra of the *Stadtumbau Ost* programme, as any concept is condemned to failure from the outset without their co-operation. In terms of urban planning policy this was breaking new ground. For, while the need for demolition could hardly be questioned in 2001, it was not yet clear how the demolition of housing could be integrated into a meaningful urban planning concept. Obviously, demolition at the whim of property owners is not the same thing as 'sustainable urban development'. Likewise, sustainable urban development cannot take place in a capitalist property market unless the market itself is functioning.

In order to implement the linkage of demolition, restructuring and upgrading measures and to ensure that planned demolition fitted in with the local urban structure, the Federal Government and the *Länder* tied the allocation of subsidies to an obligation to produce 'Integrated Urban Development Concepts'. The aim was to "link the individual measures of urban restructuring to a viable, sensible whole" (BMVBW 2001) and thus to utilise the 'opportunity' to create higher urban quality through the sensible planning of demolition and subsequent uses.

Thus high expectations were associated with *Stadtumbau Ost*, expectations which were to lead to many problems even in the planning phase. The municipal planning authorities were thereby required to produce new demographic forecasts within the brief period of less than one year in the 'Integrated Urban Development Concepts', but also had to link these with an estimate of expected economic development, an infrastructural study and thoughts on the regional context. At the same time the analysis was to be specific and detailed enough to enable the derivation of planning, implementation and financing concepts. Moreover, it was to be developed in consultation with the most important property owners, in co-operation with communities in the surrounding area and with the involvement of the citizens. To put it crudely, the municipalities were given the task of producing in a very short time a totally new type of planning concept with inadequate resources, uncertain forecasts and insufficient information and without any certainty as to their chances of implementation. This planning depended on the co-operation of interest groups (housing companies, communities in the surrounding area) and was intended to produce a consensus that would enable 'sustainable urban restructuring'.

In practice such complex analytical requirements could rarely be met in their entirety. Indeed, there were many forecasting and planning problems, so that the 'Integrated Urban Development Concepts' which were originally intended as a goal orientation are now only treated by the relevant Ministry as important "introductions to the issue". After only a short time the requirements for a new orientation of urban development planning necessary for the subsidisation of demolition were considerably reduced. This development continued in the last six years. Today, the rapid removal of surplus housing from the market dominates subsidy policy. Especially in the East German *Länder* Saxony and Saxony-Anhalt the lion's share of the available funds is spent on simple demolition. There are hardly any funds available for upgrading (e.g. such as the redevelopment of schools and kindergartens or the use of sites made available through demolition; BBR 2004). As a result *Stadtumbau Ost* is becoming increasingly less ambitious; projects for the redevelopment of empty buildings can rarely be realised.

4. Unequal co-operations

As well as the problematic aims of the programme, the inclusion of property owners who literally have a power of veto in *Stadtumbau Ost* is also frequently a significant impediment. As in theory it would be best for every individual property owner if all the other owners took their properties off the market until there was enough demand for their own empty properties, there is a classical 'fare dodger' problem: Because on the one hand all market participants benefit from the removal of surplus properties, while on the other hand the costs are only borne by those who themselves carry out this reduction, the 'fare dodger' comes out on top. As this is however true for everybody, a general blockade should occur, which can be described analogous to the game of Mikado: The first person to move loses (Bernt 2005).

The fact that under these circumstances any housing companies are willing to demolish their buildings is largely due to the subsidisation of demolition by the *Stad-*

tumbau Ost programme. It not only makes it possible to demolish buildings without making a loss, but also sweetens the owners' willingness to demolish with 'upgrading funds'. Also of major significance is the amendment of the 'Old Debts Assistance Law' (*Altschuldenhilfegesetz*) in 2001, which releases housing companies from the former German Democratic Republic (GDR) from old debts in return for demolishing housing.[1] In return for readiness to destroy their own property these companies have the opportunity to free themselves from a burden of debt that is transferred to funds for the repayment of negative inheritance and thus imposed on the general state budget. Willingness to demolish is therefore concentrated on municipal and co-operative housing companies that do not have sufficient 'mass' to cope with such losses but that are practically the only group of owners to profit from the demolition of surplus housing through the waiving of old debts. In Saxony, the leader in terms of demolition, almost 90 % of all housing demolition between 2000 and 2004 was carried out by municipal housing companies and housing co-operatives (Sächsische Aufbaubank 2005:22). All other types of owners participate hardly at all in demolition, but profit directly or indirectly from the rehabilitation of the housing market.[2]
The inclusion of different groups of owners in *Stadtumbau Ost* is also systematically unequal. As a consequence of this inequality there is a specific distribution of demolition within towns and cities, which is not primarily oriented in accordance with urban design criteria but is dependent on accessibility and availability. The result is a concentration of demolition on prefabricated slab housing (*Plattenbauten*), which is mostly administered by municipal and co-operative housing companies. Neighbourhoods of prefabricated housing, which were erected on a large scale in almost all towns and cities of the GDR from the 1970s on as a solution to the housing shortage of the time, thus become 'donor districts' which must bear the burden of the rehabilitation of the housing market on behalf of the rest of the city. The fact that development perspectives for the prefabricated housing districts, which still account for about a fifth of housing in East Germany, are increasingly disappearing from view, is only one of the problems associated with this development.
Developments in the so-called 'restructuring zones' with old buildings also give cause for concern. Demolition is in full swing here also, sometimes even listed buildings fall victim to the excavators. As a rule these are unoccupied buildings with little chance of renovation because of the poor market situation. This often affects inner city late nineteenth century housing, where it is very difficult to find tenants because

1 The so-called 'old debts' result from long-term loans made available to housing combines in the GDR for building new housing. After 1990 the relevant mortgage agreements were taken over by private banks and thus transformed into 'real' bank debts. 'Old debts' are a problem that solely affects municipal and co-operative housing companies in East Germany.

2 Where urban restructuring also involves private property, as in Leipzig or Chemnitz, it is usually limited to ruinous buildings where it is highly unlikely that anyone would be willing to fund renovation.

of traffic noise levels. Urban design planning is almost always a problem in these cases. As a result continuous street lines are broken up and the compact structure of old city neighbourhoods is damaged.

The reason for these developments is also to be found in the political mistakes made in the 1990s in the course of the transformation of the former GDR's housing system. Restitution and the commitment of the 'Old Debts Law' to privatisation mean that it is the rule today that in East German urban districts every house belongs to a different company, with headquarters spread all over Germany. Even in prefabricated districts such as Leipzig-Grünau or Halle-Neustadt, which are homogenous in comparison with old neighbourhoods in the city centres, one or two dozen different owners can be found. As all of these have varying interests because of their company form, policy and financial situation *Stadtumbau Ost* is confronted with a marked heterogeneity of interests. The large numbers of interested parties make comprehensive planning and direction extremely difficult. Because of this, demolition has hitherto been limited to owners willing to co-operate, with the result that the urban restructuring measures can scarcely be integrated into an urban design concept and in many cases themselves 'perforate' the urban structure. For example, anyone driving through certain parts of eastern Leipzig or the western periphery of Leipzig-Grünau will find a juxtaposition of carefully restored houses, crumbling ruins, elaborately designed mini parks and decaying empty sites, where little spatial logic can be discerned. There is little evidence of 'sustainable urban restructuring' here.

5. Urban restructuring and infrastructure

Added to these problems is a marked reduction of the urban restructuring programme to the problems of the housing companies involved in demolition. Looking at the debates of the last few years it can hardly be denied that water and heat provision, public transport, the development of the immediate surroundings of housing and open spaces, provision for senior citizens and residents' participation play almost no role in comparison with demolition subsidies, mortgages and old debts in the ranking of issues.

In particular the adaptation of urban public works is an area that has been much discussed but where few solutions have been implemented. The decline in the consumption of water, energy and district heating is causing growing management and technical problems for the providers. It is very difficult to adapt the networks to changing consumption. Indeed, they often have to be maintained in their entirety, even when consumption is sinking. Lower usage of the infrastructure not only leads to a poorer relation between costs and turnover, but also to technical and hygienic problems. The reduction in the through flow of drinking water and the increase in the time water spends in the network endanger the quality of the water and lead to a risk of spreading germs. In the case of waste water, lower outflow speeds lead to deposits and blockages. Thus the falling number of users not only negatively affects the general economic profitability of infrastructure providers but also leads directly to a functional collapse of the supply grids unless immense investment is poured into the systems from a certain point onwards. In order to prevent the infrastructural systems from collapsing, networks have to be reduced, redesigned and re-laid (BERNT & NAUMANN 2006).

Furthermore, the large scale demolition of housing jeopardises the water networks which in areas of prefabricated housing are usually based on collective drains. In contrast to the older urban areas, in these districts the infrastructural systems did not develop gradually but were created economically, in a centralised way and at the least possible cost (HERZ et al. 2002). The new neighbourhoods were 'hitched up' to the main lines of the supply grids for district heating and water, and the dimensions of these main lines were designed in accordance with the planned number of users. An unusual feature of these networks is the use of 'network corridors', whereby networks are not laid in the ground but put through the cellars of buildings. Thus the entire infrastructure and the buildings themselves are part of a chain, thus creating a considerable obstruction for restructuring (ibid:8). Infrastructure which was effective and modern at the time these neighbourhoods were planned thus becomes a 'trap' placing considerable restrictions on the reduction of theses districts. While the selective demolition of individual apartment blocks can contribute to the upgrading of neighbourhoods and make them less dense, this thinning out causes extra costs and expenditure for infrastructure providers. From the point of view of infrastructural economics there is a 'compulsion' to follow spatially extensive demolition strategies.

Although the adaption of urban infrastructure in the framework of the upgrading part of the urban restructuring programme is among the measures eligible for subsidies, in fact only a very small amount of funding has been used for the reduction of urban public works up to now. So in practice, little remains of the integrative and sustainable claims with which *Stadtumbau Ost* was initiated. Beyond the housing surplus a further 'mountain of problems' is already foreseeable which will require new state subsidies in a very short time.

6. Quo vadis *Stadtumbau Ost*?

The balance of urban restructuring after six years can only be described as sobering. Although it is at the core of the subsidies, the repair of market functions through the demolition of surplus provision has hitherto only worked to a very limited extent. The population continues to decline in East Germany, and as a result the housing companies continue to lose more tenants than they can compensate for by demolishing homes. A report of the *Sächsische Aufbaubank* calculates that the housing companies in Saxony must on average take four apartments off the market in order to reduce vacancies by one apartment (Sächsische Aufbaubank 2005: 13). Given this situation demolition can at best be viewed as an 'end of pipe' strategy to prevent a further escalation of the problems on the housing market. In the face of the expected continuing decline in population it is foreseeable that East Germany will have to live with continuing high vacancy levels for a very long time – and it is questionable whether it is worthwhile pushing these levels down by a few percent with such high expenditure. The plan to consolidate the housing market with the aid of publicly subsidised demolition to the extent that it an function alone again can already be described as illusory, at least in regions such as Lusatia in southern Brandenburg and eastern Saxony, the Altmark in Saxony-Anhalt or the (former) central German chemical triangle between Halle, Bitterfeld and Schkeuditz.

The sustainability of the measures already implemented is also doubtful. Under the aegis of *Stadtumbau Ost*, the demolition of empty housing currently dominates; other aspects of urban development and planning are largely neglected. There is still a lack of ideas for addressing the plight of East German towns and cities, which is due not only to a surplus of housing but also and especially to the eroded economic base, a lack of perspectives for the future and the combination of economic, infrastructural and social problems.

Although theoretically *Stadtumbau Ost* provides an opportunity to plan citywide development in a completely new way, in practice little use has been made of this opportunity so far. In the long term it is necessary to do more than merely demolish; however, this will not be possible without considerably higher funding and a political will for more planning, more direction and more state intervention.

References

Bernt, M. (2005): Stadtumbau im Gefangenendilemma. In: Weiske, C., Kabisch, S., & C. Hannemann (eds.): Kommunikative Steuerung des Stadtumbaus. Interessengegensätze, temporäre Koalitionen und Entscheidungsstrukturen in schrumpfenden Städten. Wiesbaden: 109-131.

Bernt, M. & M. Naumann (2006): Wenn der Hahn zubleibt. Wasserversorgung in schrumpfenden Städten. In: Frank, S. & M. Gandy (eds.): Hydropolis. Wasser und die Stadt der Moderne. Frankfurt am Main: 210-228.

BMVBW – Bundesministerium für Verkehr, Bau- und Wohnungswesen (2001): Wettbewerb Stadtumbau Ost. Für lebenswerte Städte und attraktives Wohnen. Auslobung und Wettbewerbsunterlagen. Berlin.

BMVBW – BUNDESMINISTERIUM FÜR VERKEHR, BAU- UND WOHNUNGSWESEN (ed.) (2005): Nachhaltige Stadtentwicklung – ein Gemeinschaftswerk. Städtebaulicher Bericht der Bundesregierung 2004. Bonn.

BMVBW – BUNDESMINISTERIUM FÜR VERKEHR, BAU- UND WOHNUNGSWESEN/BBR – BUNDESAMT FÜR BAUWESEN UND RAUMORDNUNG (eds.) (2003): Dokumentation zum Bundeswettbewerb „Stadtumbau Ost“ – für lebenswerte Städte und attraktives Wohnen. Berlin.

ENGLER, W. (2002): Die Ostdeutschen als Avantgarde. Berlin.

FUHRICH, M., & R. KALTENBRUNNER (2005): Der Osten – jetzt auch im Westen? Stadtumbau-West und Stadtumbau-Ost – zwei ungleiche Geschwister. Berliner Debatte Initial 16(6): 41-54.

HERZ, R. K. et al. (2002): Erfordernisse und Finanzierung der Anpassung der stadttechnischen Infrastruktur im Zuge des Stadtumbaus. Abschlussbericht Unterirdischer Städtebau. Dresden.

KAUFMANN, F.-X. (2005): Schrumpfende Gesellschaft. Vom Bevölkerungsrückgang und seinen Folgen. Frankfurt am Main.

KOMMISSION „WOHNUNGSWIRTSCHAFTLICHER STRUKTURWANDEL“ (2000): Wohnungswirtschaftlicher Strukturwandel in den neuen Bundesländern. Bericht der Kommission im Auftrag des Bundesministeriums für Verkehr, Bau- und Wohnungswesen. Berlin.

SÄCHSISCHE AUFBAUBANK (2005): Wohnungsmarktmonitoring 2004/2005. Dresden.

DENSITY AS A TOOL TO GUIDE URBAN SHRINKAGE CONCERNING PUBLIC WORKS?

Christiane Westphal

Density regulations are a planning tool with a long history. Urban densities are related to a multitude of important aspects of quality of life such as housing quality, quality of urban ecology and the supply of infrastructures. Therefore, density targets are set with a variety of criteria. Declining densities, as experienced in many Eastern German cities, affect the cost effectiveness, operability and environmental performance of public works such as water supply, sewage disposal and heating. Analysis of these impacts allows determination of minimum density threshold values, which differ depending on characteristics of urban form. However, definition of minimum density thresholds presents many difficulties, as cause-effect relations are not always clear. Furthermore, target parameters derived from analysing the impacts of declining densities on public works need to be synchronised with those resulting from other fields of urban planning. Despite these problems, cities with higher densities can benefit from less outlay to provide and maintain public works. Therefore, when closely interrelated to target parameters from other fields of planning, minimum densities for public works provide a useful benchmark for avoiding unmanageable density losses, thus avoiding cites with high infrastructural demands.

1. Introduction

In urban planning, density is a measurement that demonstrates the intensity of land use by calculating the ratio of quantities of use (e. g. m^2 of floor space, number of inhabitants) over a confined urban area (e. g. a hectare [ha]) (KORDA 2005). Densities of buildings, such as the floor space ratio, describe the state of the physical structure. Densities of population, e. g. net population density per hectare, indicate the potential for social interaction and customers for shopping facilities, public transport, social and technical infrastructures and so on. Indices of density are important indicators for describing aspects of quality of life in urban areas and for guiding the future intensity of land use, especially at neighbourhood level.

Urban structure densities are related to a multitude of planning issues, such as quality of housing and environment, and supply of transport, social and technical infrastructures (CHURCHMAN 1999). These interrelations are further discussed in Section 2 of this text, stressing aspects of housing quality and urban ecology. Currently, due to parallel losses of population and expansions of urbanised areas, Eastern German cities are experiencing a tremendous decline of urban densities, as outlined in Section 3. Above all, these declines in density impair the cost effectiveness, functionality and environmental performance of grid-bound public works such as water supply, sewage disposal and heating, as explained in Section 4. To limit adverse effects of declining density on public works, minimum densities need to be safeguarded. Therefore, Section 5 discusses the opportunities and constraints of defining minimum densities with regard to public works. Finally, Section 6 concludes with a discussion of the extent to which the definition of adequate or minimum densities can be used as a tool to guide urban shrinkage.

2. Density as a multidimensional urban form indicator

Density includes complex information about urban structures and is interconnected with a multitude of urban development aspects. Within urban planning, a debate about the pros and cons of higher versus lower densities has been happening for a long time (Albers 1964, Churchman 1999, Gassner 1978, Göderitz, Rainer & Hoffmann 1957, Mönninger 1994, Sieverts 1997a). Different points of view on adequate urban densities are put forward, depending on whether the focus is on housing, urban ecology, infrastructure supply or other aspects of urban planning. In Germany, for instance, these range from floor space ratios between 0.4 (Sieverts 1997b) and 4.0 (Mönninger 1994), corresponding to population densities from 80 to 800 inhabitants per net hectare.[1]

Density and housing quality

Densities vary according to different types of housing. The photographs in Fig. 1 show two examples of urban structure, with their typical densities: low density single family homes and high density prefabricated housing estates (for the densities of different types of housing, see e. g. Korda 2005, Neufert 2005).

In addition to these different densities, the housing types also provide different living conditions: single family homes offer a comparatively large proportion of private garden but hardly any public green and open space. Prefabricated housing blocks

1 The relationship between building and population density is determined by the occupation of habitable floor space per person. In 2002 in Germany the average occupation of habitable floor space was 41.6 m² per person (Statistisches Bundesamt 2004). The ratio of residential floor space to total floor space is assumed as 1:1.2. Accordingly, net population densities can be calculated using the following equation:

$$\text{Net population density in inhabitants per ha} = \frac{\text{Floor space ratio} \times 10{,}000}{\text{Habitable floor space per person} \times 1.2}$$

Fig. 1: Densities and housing quality of different housing types (photos: IÖR)

lack private open space and provide some public open space; however, they often have a monotonous design.

While low density single family homes are preferred by many residents, the supply of shopping facilities, social and transport infrastructure that also contribute to a good housing quality requires higher densities.

Therefore, target densities established to optimise housing quality vary. For suburban housing forms in a concentrated arrangement, however, floor space ratios range from 0.4 to 0.6 (corresponding to population densities from 80–120 inhabitants per net ha) (SIEVERTS 1997b). For urban housing forms, densities of 140 to 185 inhabitants per gross ha are supported (corresponding to 200–400 inhabitants per net ha and floor space ratios from 1.0–2.0)[2] (FELDTKELLER 2001).

Density and Urban Ecology

Densities of building and of population are also closely interrelated with important aspects of urban ecology, as illustrated in Fig. 2.

Firstly, the density of building, defined by the plot ratio,[3] directly influences the proportion of soil sealing, which is important for many urban ecological issues such as urban warming, surface run-off, infiltration, and recharging of ground water. For example, dense urban areas suffer from urban warming with surface temperatures 2–8° higher than those in areas of very low densities. With a 10 % increase in soil

2 Assuming that the proportion of net residential area equals 70 % of gross residential area.

3 The plot ratio determines the amount of built area in relation to the plot, e.g. a plot ratio of 0.4 means that 40 % of the plot are built on.

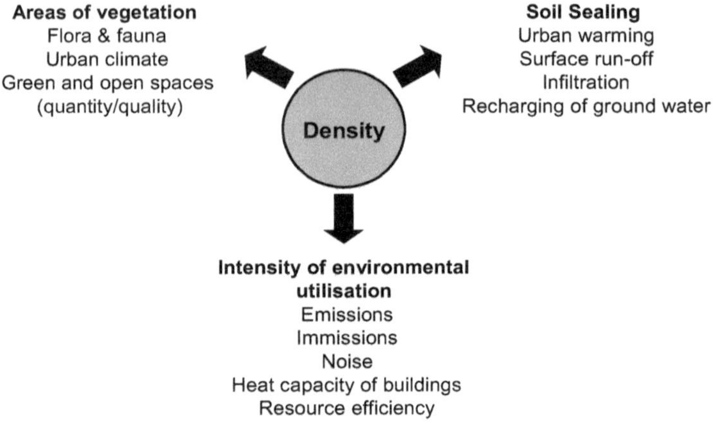

Fig. 2: Density and urban ecology

sealing, air temperature rises by 1°. Infiltration of rainfall decreases by 5 % with an increase of building density by 10 % (Pauleit 1998).

Complementary to the degree of soil sealing, the plot ratio of buildings determines possible areas of vegetation. This in turn influences development potential for flora and fauna, the urban climate, and the quality and quantity of urban green spaces. Potential for safeguarding areas of vegetation decline as densities increase (Pauleit 1998, Sukopp & Wittig 1998).

Furthermore, building and population densities are closely interrelated with the intensity of environmental use, e. g. increasing densities lead to growing amounts of emissions and nuisance, as well as to an increase of the heat capacity of buildings. Apart from these predominantly negative effects of higher densities on urban ecology, densification also brings about positive effects, as residential areas of higher densities are generally more resource efficient since they have lower heat requirements, cause less traffic, and so on (Owens 1986).

Evaluation of the pros and cons of lower versus higher densities led Happe et al. (1994) to set some target values for densities derived from the study of urban ecology for different types of urban areas. While target densities for city centre and mixed used areas range from floor space ratios of 1.2 to 2.4, (corresponding to 240–480 inhabitants per net hectare), those for housing at the urban fringe are comparatively lower, ranging from 0.6 to 1.0. Stressing the negative effects of densification on the protection of soils as well as on urban climate, Losch (1992) suggests limiting densities to a maximum floor space ratio of 0.7 and Sieverts (1997b) recommends maximum floor space ratios of 0.4 to 0.6 (corresponding to 80–120 inhabitants per net hectare).

3. Density loss in Eastern Germany

Urban shrinkage, as experienced in Eastern Germany, leads to a decline of urban densities, including population and buildings. The city of Leipzig, for instance, suffered a decline in net population densities from 148 inhabitants per gross hectare of residential area in 1985 to 76 in 2001 (Büro für urbane Projekte 2004). However, density loss does not occur equally in all parts of a city: Eastern German cities experience parallel processes of growing and declining densities in neighbouring urban areas. This leads to a wide variety of densities for physical structures and population, including high density developments, widespread vacancies, and the evolvement of urban wastelands. Due to the high vacancy rates of flats, densities of physical structures and densities of population often become contradictory. For example, physically dense multi-storey housing with vacancy rate of 50 %, shows population densities corresponding to those of physically much less dense single family homes (Schiller & Siedentop 2005).

Various concepts of adequate densities in growing cities have been developed in the context of the debate on urban planning. Currently, two contrasting concepts are discussed: dense and compact European cities with desired floor space ratios from 0.7 to 4.0 (Fuhrich 2004, Mönninger 1994), and the disperse settlement structures such as *"Zwischenstadt"* or *"Netzstadt"* aiming at floor space ratios of between 0.4 and 0.6 (Sieverts 1997b).

However, ideas about adequate densities for shrinking cities are still lacking. So far, some models of urban structural change for shrinking cities have been discussed: 'contraction' to a compact urban core, 'perforation', (a gradual erosion of urban structures), or 'dispersion', (a parallel erosion and extension of settlements that leads to an extreme reduction of urban densities. Quantitative goals for adequate densities in shrinking cities have not yet been defined. Nonetheless, concepts and guidelines on adequate densities are needed to safeguard functional and attractive urban cores in shrinking cities, especially with regard to public works, as discussed in the following section.

4. Urban shrinkage, density loss and public works

Interrelations of density and public works

Population and building densities are closely interlinked with the cost effectiveness and environmental performance of grid-bound public works such as water supply,

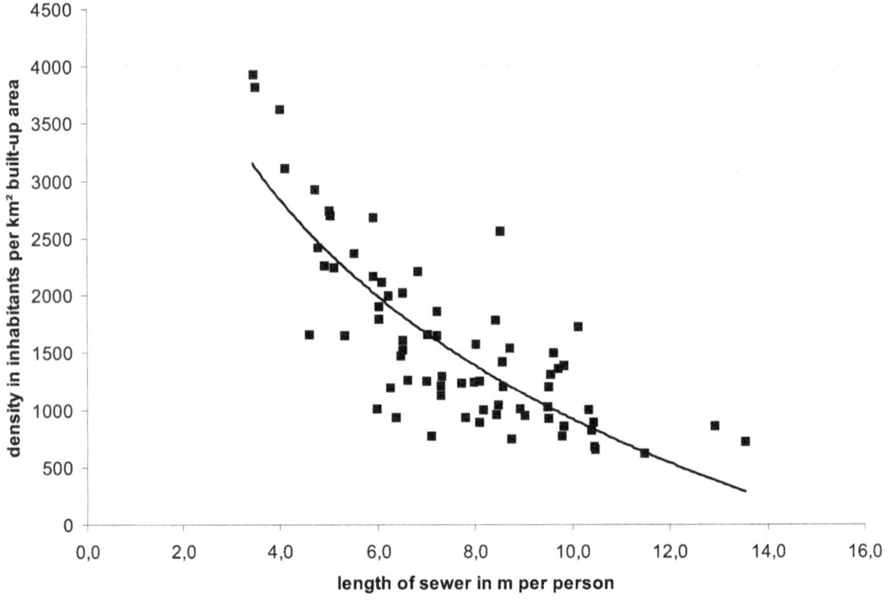

Fig. 3: Density and length of sewers per capita in 2004 (calculation based on data from the statistical offices of Sachsen, Sachsen-Anhalt and Brandenburg)

sewage disposal and district heating. Water supply and sewage disposal relate to use and quantity per capita and therefore to population density; heat supply relates to heatable dwelling space and therefore to building densities.

In general, higher densities are connected with cost effectiveness and environmental sustainability, while lower and declining densities lead to higher per capita costs, problems of operability and increasing resource consumption (BUCHERT et al. 2004, SCHILLER & SIEDENTOP 2005, TIETZ 2006). For example, Fig. 3 shows a scatter plot of population densities compared to length of sewers per capita. Each point represents values for an administrative district (corresponding to NUTS 3). The diagram illustrates the exponential increase of sewer lengths in areas with decreasing densities. While densities of more than 3,000 inhabitants/km² of built-up area correspond to pipe lengths of about 4 m per capita, densities of 1,500 inhabitants/km² correspond to pipe length of 6 to 10 m per capita, and densities of less than 1,000 inhabitants/km² relate to pipe lengths of 8 to 14 m.

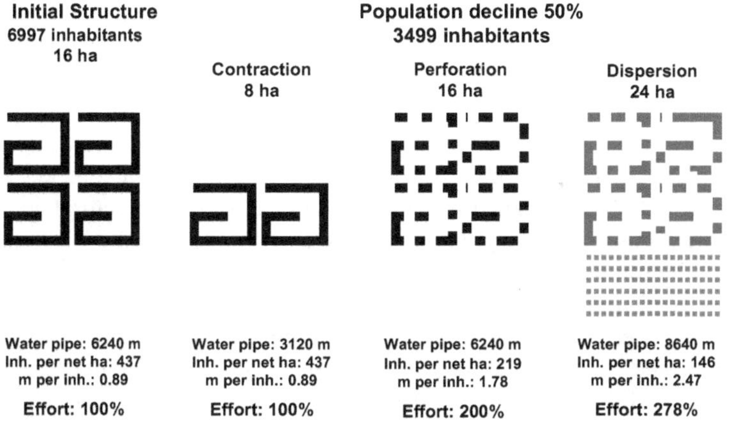

Fig. 4: Modelling infrastructural efforts according to patterns of urban shrinkage (calculations based on BUCHERT et al. 2004)

Impacts of density loss on public works

The process of urban shrinkage, leading to a decline of densities, has a considerable impact on public works. Declining population densities lead to lower demands causing technical problems. An example is the supply of potable water; as flow velocity of water declines, the period spent within the pipes increases and causes hygiene problems. Decreased velocity of wastewater causes sedimentation, noxious odours and corrosion of pipes. Declining densities also affect the economic viability of expensive district heating distribution systems, which are only viable for high density areas. Due to the high cost of public works, a decline of population and building densities leads to an increase of per capita costs for provision and maintenance. Additional costs arise when further efforts are made to safeguard the operability of those infrastructures, for example for flushing water and wastewater pipes. Environmental performance of public works is affected by increasing material input for the provision of services per capita. Furthermore, environmentally-friendly district heating is in danger of being substituted by less environmentally friendly forms of heat supply, such as electric heating (FREUDENBERG & KOZIOL 2003, HERZ, WERNER & MARSCHKE 2002, TIETZ 2006).

Fig. 4 shows an example of modelling the effort to sustain grid-bound public works for different scenarios of shrinkage. Based on data of water pipe lengths for different areas of distinctive urban character (BUCHERT et al. 2004), the infrastructural effort, represented by the length of water pipe per capita, is modelled for a residential area of prefabricated housing blocks of 16 hectares (net residential area). The infrastructural effort of this initial situation with 0.89 m length of water pipe per capita is set as 100 %.[4] Depending on the location of demolition, reconstruction and construction of new homes on non-urbanised land, the infrastructural effort, varies from 100 % to approximately 280 %.

Consistently demolishing vacant buildings from the fringe of settlements towards their centres, described as 'contraction,' allows the retention of compact urban structures with high densities and the complete removal of infrastructural networks from their extremities. With a population decline of 50 %, half of the infrastructural network can be removed, and it is possible to more or less avoid additional efforts to sustain public works, and thus keep the infrastructural effort at approximately 100 %.

'Perforation' describes a more or less consistent decline of density over larger urban areas, leading to the demolition of single buildings throughout the urban area. Demolishing single buildings amidst remaining buildings does not, however, allow removal of infrastructural networks, so that the per capita effort to maintain these structures (indicated by pipe length per capita) increases to 200 %. In addition to this perforated demolition of single buildings, when the urbanised area is expanding by building new single family homes at the fringe of settlements – known as 'dispersion' – the length of water pipes per capita, and therefore the effort to maintain grid-bound public works, will easily expand towards 280 %, as infrastructural networks need to be extended.[5]

5. Minimum densities with regard to public works - opportunities and constraints

Examination of the relationship between density, urban shrinkage and public works leads to the question of whether it is possible to define minimum densities with regard to public works. Some threshold values may be defined by analysis of these interrelations, a literature review and calculations.

For instance, analysis of standard heat requirements for different types of settlement structures indicates that floor space ratios of 0.8 to 1.0 (corresponding to 160–200 inhabitants per net ha) are a minimum for the cost effectiveness of environmentally friendly district heat supply (Roth et al. 1980, Siedentop et al. 2006, Tietz 2006).

4 For the area of prefabricated housing blocks the floor space ratio is assumed at 1.5, the length of water pipes per m² floor space as 0.026 m and the floor space per person as 34.3 m (2.1 inhabitants in a flat of 60 m², ratio of dwelling floor space to floor space 1:1.2) (based on Buchert et al. 2004; Siedentop et al. 2006).

5 For the additional area of single family homes a floor space ratio of 0.4 is assumed as well as a floor space of 47.5 m² per person (2.3 inhabitants in a flat of 91 m², ratio of dwelling floor space to floor space 1:1.2). Water pipes' length per m² floor space is set as 0.075 m/m² (Buchert et al. 2004, Siedentop et al. 2006). Within this model it is expected that 20 % of the population move into these new single family homes.

As the effort to provide and maintain public works grows exponentially when densities decline, threshold values can also be defined by determining the density value, the surpassing of which leads to a significant increase of effort, e. g. of pipe length per capita. This density value differs according to specific urban characters. While for single family homes infrastructural efforts greatly increase when the net population density falls below 30 inhabitants per net hectare (floor space ratio of 0.15), for housing from the turn of the twentieth century demands increase when the net population density decreases to below 160 inhabitants per net hectare (floor space ratio 0.8) (calculations based on Siedentop et al. 2006).

However, using public works to define quantitative threshold values for urban densities presents many problems. Although density influences the efficiency of public works this efficiency also depends on many other factors such as location, topography or technical characteristics (Gassner & Thünker 1992). For example, technical problems with water supply relate to declining population as well as decreased water consumption per capita (e. g. from about 200 litres per day, per person in Eastern Germany during the 1980s to 90 litres in 2005) (Freudenberg & Koziol 2003). Yet, the influence of these factors cannot always be clearly identified.

Urban restructuring costs are not solely caused by declining urban densities but are also influenced by the sequence of demolishing vacant buildings and restructuring urban areas. Effective cooperation between stakeholders ensures successful coordination of the process of urban restructuring, thereby avoiding additional costs for expensive interim solutions (Lindner & Buhtz 2006).

Moreover, setting minimum densities correlates to current societal context. Therefore, it depends on normative and political decisions, e. g. maximum prices or resource consumption for supply of infrastructure. Other hurdles relate to the limited availability of data, especially as operators of public works restrict its availabilty.

6. Conclusion: Density as a tool to guide urban shrinkage?

Density is just one of several important planning parameters. The management of shrinkage requires consideration of many aspects, such as norms and political preferences, process sequences, locations of urban restructuring areas, and many other factors. Minimum density values cannot be set across the board, but should be calculated for particular urban characters (e. g. single family homes, turn of the twentieth century housing, prefabricated housing blocks).

Nevertheless, minimum density concepts are a useful tool for guiding the process of urban shrinkage. With all other conditions being equal, dense cities require less effort to provide and maintain public works. Uncontrolled, dispersed shrinkage and density losses will undoubtedly lead to overwhelming infrastructural demands. From the point of view of minimising infrastructural efforts and safeguarding adequate infrastructural supplies in areas of single family homes, density should not decline below 30 inhabitants per net hectare (floor space ratio 0.15). In denser urban structures, such as prefabricated slab housing or housing from the early twentieth century, density should not decline below 160 inhabitants per net hectare (floor space ratio 0.8). However, as density is also closely related to many other issues of urban planning, minimum densities cannot be defined simply according to the needs of public works. As outlined above, contrary to the needs of public works, higher densities adversely affect high quality urban ecology. Adequate densities for housing at the urban fringe that ensure urban ecological quality range from floor space ratios of 0.4 to 1.0 (corresponding to 80–200 inhabitants per net hectare). For urban structures of higher densities these target values range from floor space ratios of 1.2 to 2.4 (corresponding to 240–480 inhabitants per net hectare). Regarding housing quality, density targets range from floor space ratios of 0.4 for suburban housing forms to 2.0 for urban housing (80–400 inhabitants).

Comparison of density targets resulting from calculating infrastructural demand in shrinking cities with density targets from other fields of urban planning reveals that, in spite of the significant effects of declining densities on public works, density targets established for this field of urban planning seem to be lower than target parameters from other fields of urban planning. However, due to the inflexible grid-bound logic of public works, density targets defined from this perspective in shrinking cities appear to be much more strict.

Finally, it can be concluded, that, although being a difficult task, defining minimum or adequate densities for shrinking cities is of a great use to guide urban shrinkage, when various fields of planning combined with varying densities in areas of distinctive urban characters are taken into account. As a result of defining these target densities, city-wide density models mapping the target values (Westphal & Hutter 2006) might be developed.

References

Albers, G. (1964): Wohndichte und Geschoßflächenzahl. StadtBauwelt, 1/1964: 44-48.

Buchert, M., Fritsche, U., Jenseit, W., Rausch, L., Deilmann, C., Schiller, G., Siedentop, S. & A. Lipkow (2004): Nachhaltiges Bauen und Wohnen in Deutschland, Stoffflussbezogene Bausteine für ein nationales Konzept der nachhaltigen Entwicklung - Verknüpfung des Bereiches Bauen und Wohnen mit dem komplementären Bereich „Öffentliche Infrastruktur“. Umweltbundesamt: UBA Texte 01/04. Berlin.

Büro für urbane Projekte (2004): GIS-Modell, Verortung. In: Stadt Leipzig, empirica, Büro für urbane Projekte (eds.): Leipzig 2030. Forschungsbericht im Rahmen des Ideenwettbewerbs Stadt 2030 gefördert vom Bundesministerium für Bildung und Forschung. Leipzig: 57-84.

Churchman, A. (1999): Disentangling the Concept of Density. Journal of Planning Literature 3 (4): 389-411.

Feldtkeller, A., (ed.) (2001): Städtebau: Vielfalt und Integration, Neue Konzepte für den Umgang mit Stadtbrachen. Stuttgart, München.

Freudenberg, D. & M. Koziol (2003): Anpassung der technischen Infrastruktur beim Stadtumbau, Arbeitshilfe. ISW-Schriftenreihe: Fachbeiträge zu Stadtentwicklung und Wohnen im Land Brandenburg 2/2003. Frankfurt (O.).

Fuhrich, M. (2004): Kapitel III: Innenentwicklung vor Außenentwicklung - Erfahrungen aus dem Forschungsfeld „Städte der Zukunft". In: BBR (Bundesamt für Bauwesen und Raumordnung) (ed.): Bauland- und Immobilienmarktberichte - Ausgabe 2004, Band 19. Bonn: 81-93.

Gassner, E. (1978): Die Grenzen der Verdichtung bei Wohnbaugebieten. In: Borchard, K. & E. Weiss (eds.): Aus Verantwortung für die Gestaltung unserer Umwelt, ausgewählte Beiträge von Edmund Gassner. Schriftenreihe des Instituts für Städtebau, Bodenordnung und Kulturtechnik der Universität Bonn: Beiträge zu Städtebau und Bodenordnung 12. Bonn: 93-119.

Gassner, E. & H. Thünker (1992): Die technische Infrastruktur in der Bauleitplanung, 2nd ed. Berlin.

Göderitz, J., Rainer, R. & H. Hoffmann (1957): Die gegliederte und aufgelockerte Stadt. Archiv für Städtebau und Landesplanung. Tübingen.

Happe, M., Hinzen, A., Pieper, H., Davids, P. & F. Terfrüchte (1994): Grünbuch Planung - Bausteine für die Planungspraxis in Nordrhein-Westfalen: Institut für Landes- und Stadtentwicklungsforschung des Landes Nordrhein-Westfalen (ILS) im Auftrag des Ministeriums für Stadtentwicklung und Verkehr des Landes Nordrheinwestfalen (MSV). Dortmund.

Herz, R., Werner, M. & L. Marschke (2002): Anpassung der technischen Infrastruktur. In: Bundesministerium für Verkehr, Bau- und Wohnungswesen (ed.): Fachdokumentation zum Bundeswettbewerb „Stadtumbau Ost" - Expertisen zu städtebaulichen und wohnungswirtschaftlichen Aspekten des Stadtumbaus. Bonn: 50-60.

Korda, M., (ed.) (2005): Städtebau, Technische Grundlagen. 5th ed. Stuttgart, Leipzig.

Lindner, M. & M. Buhtz (2006): Technische Infrastruktur bei Abriss. BundesBauBlatt 11/2006: 16-20.

Losch, S. (1992): Sparsame und schonende Flächeninanspruchnahme - ein unerfülltes Ziel? Zeitschrift für angewandte Umweltforschung 5 (1): 90-102.

MÖNNINGER, M. (1994): Die Angst vor der Dichte, Städtebau in den neunziger Jahren. In: Dezernat Planung der Stadt Frankfurt am Main (ed.): Region. Die Zukunft des Städtischen: Frankfurter Beiträge Band 5. Frankfurt, New York: 163-168.

NEUFERT, E. (2005): Neufert Bauentwurfslehre, 38th ed. Wiesbaden.

OWENS, S. (2006): Energy, planning and urban form. London.

PAULEIT, S. (1998): Das Umweltwirkgefüge städtischer Siedlungsstrukturen, Darstellung des städtischen Ökosystems durch eine Strukturtypenkartierung zur Bestimmung von Umweltqualitätszielen für die Stadtplanung. Landschaftsökologie Weihenstephan 12. Freising.

ROTH, U., HÄUBI, F., ALBRECHT, J., BISCHOFF, M., DEUCHER, A., HARDER, L., LANGRAF, B. & G. PAPE (1980): Wechselwirkungen zwischen der Siedlungsstruktur und Wärmeversorgungssystemen. Bonn.

SCHILLER, G. & S. SIEDENTOP (2005): Infrastrukturfolgekosten der Siedlungsentwicklung unter Schrumpfungsbedingungen. DISP 160: 83-93.

SIEDENTOP, S., SCHILLER, G., KOZIOL, M., WALTHER, J. & J.-M. GUTSCHE (2006): Siedlungsentwicklung und Infrastrukturfolgekosten - Bilanzierung und Strategieentwicklung, Endbericht. Bonn.

SIEVERTS, T. (1997a): Einige Anmerkungen zum Thema „Dichte". In: Magistrat der Stadt Frankfurt am Main: Dezernat Planung - Amt für kommunale Gesamtentwicklung und Stadtplanung (ed.): Planung und Entwicklung neuer Stadtteile, Internationaler Frankfurter Städtebau-Diskurs, Deutsches Architektur-Museum, 22.-23. Mai 1997. Frankfurt am Main: 83-86.

SIEVERTS, T. (1997b): Zwischenstadt - zwischen Ort und Welt, Raum und Zeit, Stadt und Land. Braunschweig, Wiesbaden.

STATISTISCHES BUNDESAMT (2004): Bautätigkeit und Wohnungen, Mikrozensus-Zusatzerhebung 2002, Bestand und Struktur der Wohneinheiten, Wohnsituation der Haushalte, Fachserie 5 (1). Wiesbaden.

SUKOPP, H. & R. WITTIG (eds.) (1998): Stadtökologie, Ein Fachbuch für Studium und Praxis, 2nd ed. Stuttgart, Jena.

TIETZ, H.-P. (2006): Auswirkungen des demographischen Wandels auf die Netzinfrastruktur. In: GANS. P. & A. SCHMITZ-VELTIN (eds.): Demographische Trends in Deutschland – Folgen für Städte und Regionen, Akademie für Raumforschung und Landesplanung (ARL): Räumliche Konsequenzen des demographischen Wandels Teil 6. Hannover: 155-171.

WESTPHAL, C. & G. HUTTER (2006): Dichtemodelle und ihre Integration in kommunale Strategien für eine qualitative Innenentwicklung. In: GENSKE, D., HUCH, M. & B. MÜLLER (eds.): Fläche - Zukunft - Raum. Strategien und Instrumente für Regionen im Umbruch. Schriftenreihe der Deutschen Gesellschaft für Geowissenschaften 37: 77-90.

SHRINKING CITIES – OPPORTUNITIES AND CHALLENGES FOR URBAN GREEN SPACE PLANNING: EXPERIENCES AT THE LARGE HOUSING ESTATE OF LEIPZIG-GRÜNAU

Stefanie Rößler

Spatial shrinkage in cities occurs as a result of demographic change and economic decline and produces a major challenge for future urban and green space development. Following a brief introduction to the stated phenomenon and the various ways in which it manifests itself and is dealt with in the cities of Eastern Germany, this article will discuss the changing framework of, as well as the requirements for, green space planning in shrinking cities. This discussion will be followed by a more detailed presentation of the observations and insights made at the large housing estate Leipzig-Grünau (Germany).

1. Shrinkage and its impacts on urban structure

The phenomenon of shrinkage has an impact on various aspects of urban development in diverse ways. Seen from a spatial perspective, the most pressing problems are the fall in the demand for housing, commercial and industrial property, and the closing of public institutions such as schools, thus creating a surplus of building stock. The increasing number of vacancies, in turn, has led to the demolition of buildings, producing more empty spaces and transforming the urban fabric. The demolition of buildings coupled with fewer requirements for redevelopment of those vacant lots has brought about an increase in open spaces.

Urban restructuring

Several policies and joint federal-state programmes are addressing the spatial effects of shrinking phenomena in Germany. The *Stadtumbau Ost* ('Urban Restructuring East') programme was introduced in 2002 to adjust the housing market to population loss and the resulting housing demand crisis and to enhance the attractiveness of shrinking cities. Through this programme, the government grants financial aid to real estate owners and public utilities with the aim of stabilising the housing market by

(1) demolishing vacant buildings and obsolete infrastructure and
(2) the parallel revaluation of affected quarters.[1]

1 During the period (2002-2009) approximately 2,5 billion EUR will be provided. By 2007, about 170,000 flats (28.02.2007) had been demolished in Eastern Germany by subsidies from this programme and another 200,000 flats are planned to be demolished by 2009 (BM-VBS & BBR 2007).

By claiming financial aid for demolition, the property owner is obliged not to build any new residential buildings on the site for the subsequent ten years. Thus, the urban restructuring approach of dealing with the spatial consequences of shrinking phenomena leads to changes in the urban structure, mainly apparent by the increasing number of open spaces. In accordance with these legal constraints and the low demand for new housing, green space development appears to be the only solution for reusing demolition sites as well as a strategy for improving the quality of life and the environment in affected neighbourhoods and cities. Taking into account the facts that shrinking cities also show tendencies of concurrent suburbanisation and land-use change, as well as the fact that there are fewer chances to realise inward urban development by reusing brownfield sites for commerce and housing, the spatial density of shrinking cities will be reduced.

Relevance of green spaces

Urban restructuring is generally understood as a response to the current situation of shrinkage in cities of Eastern Germany. However, it is also viewed as a means of enhancing quality of life and sustainable development in shrinking cities. Green spaces are supposed to both provide an answer to the resulting open spaces and an intrinsic part of a sustainable city. On account of this, green space planning, landscape architecture and urban ecology are important fields of action in shrinking cities. The increasing number of open spaces provides opportunities for the long-term development of green spaces: the realisation of green space systems within the urban fabric, and improvement of the residential environment and ecological situation. These demands, however, are confronted by manifold barriers in practice. Besides the impact on the built environment, shrinkage is accompanied by demographic change and economic problems. Shrinking processes and the practice of restructuring result in a fundamental change in conditions both for the design and function of future urban green spaces as well as for the realisation, organisation and funding of green space planning.

2. Challenges for green space development in shrinking cities

The opportunities for green space development[2] in shrinking cities are confronted by manifold challenges in practice. The main aspects of shrinkage in Eastern Germany and their spatial effects will be discussed in relation to their impact on the develop-

2 Green spaces are understood as open spaces within the urban fabric which are dominated by vegetation. They can be relics of the original natural landscape or deliberately planned and designed green spaces in public or private use and/or responsibility (for example gardens, urban parks, allotments).

ment of green spaces in this article. In accordance with these frameworks, the possibilities of further development will be described.

Frameworks for shrinking cities

Demographic and societal change

Demographic change and changes to lifestyles may result in altered qualitative and quantitative demands on green spaces. These, in turn, may require new forms and functions in order to foster the use of future green spaces.

(1) The local decrease in population may lead to a reduced number of users of green spaces. Depending on the specific character of a particular green space, this may result in the reduction of provision in different neighbourhoods. The fact that the behaviour of users and their perception of safety depends on the density of users must be taken into consideration. Therefore, declining numbers of users may result in problems with vandalism and safety and, ultimately, in people's reluctance to use such spaces. In densely built quarters that currently lack green spaces, the situation may be improved by means of a better provision of green spaces. The common approach ensuring an equal and equitable provision of high-quality green spaces in all parts of the urban fabric may be reconsidered, taking into consideration the potential number and types of users. Furthermore, based on the growing number of green spaces and declining budgets, it can be expected that more natural and multifunctional green spaces will be available to a larger number of users.

(2) In addition to the population decrease, societal changes and alterations in the population structure are important for future demands on green spaces (NOHL 2002). It is difficult, however, to predict whether lifestyles do, in fact, have an impact on green space requirements without a systematic analysis of the issue (TESSIN 2004). Nevertheless, demographic and societal changes resulting in population shifts in different groups of people may change demands and require specific varieties of green spaces; for example, increasing demands of green spaces near residential areas for the growing proportion of elderly people (SCHMIDT 2005) and spaces for immigrants, who may have particular demands on public and private green spaces.

Economic framework

The economic situation of shrinking cities influences the capacity of public authorities and private owners to finance and maintain green spaces. It can be safely assumed that demographic changes will limit public bodies in their capacity to fulfil their duties in many public fields and that this will, therefore, also hold true regarding the provision of public green spaces (SEITZ 2006). Besides financial constraints, the growing number of (potential) green spaces arising from the demolition of buildings may necessitate new and innovative forms of funding and management (BECKER 2003). The municipal budgets will need to be supplemented by institutional subsidies and private money. However, the effectiveness and applicability for sustainable and high-quality development of green spaces will have to be proven (PREISLER-HOLL 2004; BBR 2004). Although supported by private money, the public duty of providing accessible green spaces that are ecological as well as aesthetical needs to be ensured. Besides innovative forms of financing, new means of cooperation with private stakeholders, local economy and residents have to be established in order to relieve public authorities.

Opportunities for green space development in shrinking cities

The increase of open spaces combined with lack of pressure to rebuild on sites provides opportunities to improve the green space provision in shrinking cities unrestricted by competing uses. Transforming vacant lots into green spaces is usually the only solution for reuse and, at the same time, may represent a contribution to sustainable city development. These capacities also mean new, and at times ambiguous, demands on green spaces, lying somewhere between being an interim solution and a location factor (GISEKE 2002). The development of open and green spaces, however, means tightrope walking between the opportunity to create new green spaces and risking the loss of urbanity, thus creating a negative perception of decay (DOEHLER-BEHZADI et al. 2005). It is expected that spatial transformation processes in shrinking cities will be accompanied by a changing perception of green spaces and their relevance.

Restructuring schemes and urban models

The implementation of urban restructuring determines the future of urban structures as well as the relationship between open and built spaces to a significant extent. Shrinking cities establish distinctive models of urban development, based on their historical urban structure, specific pressures, development objectives, and legal and economic restrictions. These models also influence the distribution and shape of the open spaces resulting from demolition, which may be potential green spaces. The growing amount of open and green spaces should be regarded as an important, and sometimes even the determining, element of future urban models and concepts (e. g. REUTHER 2002). The choice of a specific urban model also highlights the importance and the value attached to green spaces within the urban restructuring process. The practice and professional discourse of urban restructuring ranges between two extremes (KÜHN 2000; HESSE & SCHMITZ 1998; BLUME 2005):

(1) A compact city, following the ideals of the European City and
(2) The acceptance and deliberate development of the dissolutive tendencies of the city in its urban region.

The political and environmental paradigm of a sustainable urban development to be achieved by avoiding suburban sprawl and enhancing the urban inward development is usually followed by the demand for compact urban structures (RNE 2004; BMVBW 2005). This influences the debate about the quantity and quality of existing and potential urban green spaces with regard to the aim of sustainability. On the one hand, a long-term network of different green spaces within the city offers chances for improving quality of life and the environment. On the other hand, open spaces and brownfield sites are potential spaces that can be rebuilt, thus fostering an inward urban development while, at the same time, fostering the preservation of open spaces in the urban region (HUTTER et al. 2004; HEILAND 2005).

Despite the ambitious objectives of urban restructuring and sustainable development, the reality of shrinking cities in Eastern Germany reveals several difficulties. Firstly, it is apparent that demolition and construction follow economic constraints and are determined more by owners and developers than by planning authorities. This results in differentiated patterns of shrinkage and land use, showing tendencies of both maintaining density and becoming fragmented. The restrictions that accompany adherence to the ideal of the European City may limit steering possibilities for developing sustainable cities under shrinking conditions. Dealing with models geared to a fragmented urban form may result in more opportunities being realised. The choice of an urban model also affects the possibilities for the development of urban green space:
(1) Patterns of concentration at one or more urban core mean a relatively dense urban structure without open spaces within the urban fabric but outside the borders or in between different urban cores. These interspaces between urban islands implicate new opportunities but also demands for new types of green spaces.
(2) The perforation of urban structures is followed more or less by the disintegration and fragmentation of the urban form, generating smaller, scattered open spaces directly in the vicinity of buildings.

Both patterns result in different sets of framework conditions that act in accordance with ecological functions, aesthetic perceptions and the possible functions of existing and potential green spaces.

Green space development in residential neighbourhoods
The implementation of both components of the approach to urban restructuring – demolition and revaluation – lead to different effects on green space development

depending on specific urban structures. There are different sets of general framework conditions and possibilities of development in quarters of the older city and at the housing complexes dating from the Socialist period.

(1) The demolition of single, vacant and derelict buildings in quarters of the older city (buildings designed in the Wilhelminian style) results in small-scale dedensification. In several ways this has had, and continues to have, a positive impact on the development of new green spaces in neighbourhoods, which so far have been poorly supplied. In areas located in the direct vicinity of residential buildings, new offers for private and public green spaces can be made. As well as the positive effects on the residential environment, the fact has to be taken into consideration that urban qualities are strongly determined by built structures, especially in the historic city. Thus, having less spatial density may result in a loss of urbanity. This also depends on the shape and function of resulting green spaces on former built plots. After all, minimum standards of design and maintenance are the precondition for a real gain in quality. However, reuse of these areas can also be problematic. Private real estate owners of single plots often want to retain their properties with the aim of speculating on their future economic value. Consequently, the interim use of building land and even its conversion to green space may result in a loss of value.

(2) The greatest proportion of demolition is currently being happening at large housing estates, which are mostly located on the outskirts of cities. Two different approaches to urban restructuring can be distinguished, depending on the distribution and quantity of vacancies, the level of renovation of the housing stock, and the structure as well as behaviour of the owners:

(a) The demolition of buildings is concentrated on the edge of affected neighbourhoods. This enables the stabilisation of remaining built structures and maintains operational conditions of the public works (for the aspects of public works see also the paper of C. Westphal in this book). Large green spaces can be developed at the edge of the remaining housing stock, serving as recreational areas and natural habitats connected to the surrounding rural landscape. For the most part, there is no pressure to rebuild these sites; thus, a policy of allowing them to revert to nature can be supported. However, conventional approaches to design and use urban green space are limited under these specific conditions, and this should be taken into consideration. To establish these new green spaces it is necessary to develop innovative types of design and forms of funding (e.g. urban forests, agricultural uses and wilderness).

Fig. 1: Reuse of a former built site – aesthetical and functional quality? (All photos taken at Leipzig-Grünau, S. RÖSSLER)

Fig. 2: Leasing open spaces to residents for private use as gardens

(b) In certain cases, on account of heterogeneous structures of owners, economic constraints and living preferences of residents, the demolition of buildings leads to a widespread scattering of housing stock. Small scale interventions distributed throughout the entire estate enlarge the existing green space system. Declining budgets complicate management and maintenance of the increasing number of green spaces. In short, it is difficult to ensure simple design and functionality. However, if there are no visible improvements there will be no perception of the intended revaluation but only one of decay (Fig. 1). Quantitative provision of public green spaces is fairly adequate in large housing estates. Thus, an increasing amount of green spaces does not necessarily mean an improvement of the situation. In such cases, several efforts have to be made in order to provide usable and accessible green spaces for a wider scope of activities (Fig. 2).

Ecological demands

Green spaces are crucial elements of the urban structure, with their variety of ecological functions and environmental benefits, provide habitats for animals support the diversity of species and mitigate the environmental stress of an urban climate. Besides the positive quantitative effects of an increasing number of open and green spaces, the quality of habitats is also determined by the connection between different green spaces, structural diversity and level of disturbance by users or traffic.

A widespread approach to deal with the increasing amount of green spaces and declining budgets involves reducing the amount of design and maintenance work done in favour of both lowering costs and fostering a more natural development (Fig. 3). Thus, the adaptation to current framework conditions as well as realisation of eco-

Fig. 3: Maintenance and more natural development at demolition sites

Fig. 4: Simply designed green space with low ecological and functional values

logical objectives may be enhanced. This approach, however, must be considered in a critical manner. If, for example, ecological solutions are simply introduced as stopgaps due to a lack of users as well as a lack of ideas related to design and financing, then ecological qualities will not increase automatically. If green spaces appear to be neglected, then the perception of the urban restructuring process may be negative. To improve the quality of life and the environment, ecological values as well as aesthetic and social qualities should be taken seriously (Fig. 4).

3. Example: Large housing estate Leipzig-Grünau (Germany)

The previously discussed factors determine the framework conditions for green space development in the context of urban spatial shrinkage. In this section, the general issues will be analysed using the example of the large housing estate Grünau in Leipzig. General information concerning the research background is followed by a short description of the example and, finally, presentation of the principal results.

The research project

The research study "Role of Green Space Planning in the Process of Urban Restructuring in Leipzig-Grünau" explored the effects, problems and challenges of urban restructuring and identified several potential ideas and specific recommendations for green space development using the example of the large housing estate Leipzig-Grünau.[3] The research focused on the integration of issues of green space develop-

3 This study was part of the research project "Sociological Observations on the Implementation and Execution of Demolitions, Aiming at the Reduction of Housing Vacancies and the Improvement of Environmental Conditions and the Quality of Life in the Large Housing Estate Leipzig-Grünau" carried out at the UFZ Leipzig (Helmholtz Centre for Environmental Research) in cooperation with the IOER Dresden (Leibniz-Institute of Ecological and

ment into town planning by evaluating relevant planning documents, frameworks, strategy papers, master plans and concrete projects and measures. This was completed by an assessment of the process of urban restructuring, focusing on issues of governance, management, cooperation and the participation of decision-makers and residents with regard to economic, legal and planning challenges.[4]

Large housing estate Leipzig-Grünau

The Grünau housing estate was built on the western border of Leipzig between 1975 and 1988. Since 1990, the estate has been supported by different political initiatives and large financial programmes for improving quality of life for residents by renovating buildings, improving infrastructure and developing public realm and green spaces. Nevertheless, Grünau has been and continues to be afflicted by shrinkage like many other cities and neighbourhoods in Eastern Germany.

The original scheme provided 36,000 flats for approximately 85,000 residents. Since 1997/98, Grünau has lost between 33 and 52 % of its population with some neighbourhoods being affected more than others. Currently, about 46,000 residents (12/2006) are living in Grünau. By 2005, about 3,800 flats had been demolished. The average vacancy rate of flats is still about 20 %, with large differences between the neighbourhoods. Different scenarios forecast between 32,000 and 40,000 residents by 2020. Therefore, a further reduction of the housing stock is necessary in order to ensure economic and operational management for the owners (Stadt Leipzig 2004; Stadt Leipzig 2005; Stadt Leipzig 2007). Besides the demographic development, the structural problems of Grünau are manifold as it features buildings of various structures, ages, types, levels of renovation and social structures in each neighbourhood. The economic situation of the residents is comparable to the city as a whole but differs among its several neighbourhoods. There is also a high proportion of elderly people (Bernt & Kabisch 2006).

In response to the challenges of the shrinking processes and the political initiatives for urban restructuring, the *Urban Development Concept*[5] was developed by the municipality in 2002 (Stadt Leipzig 2002). Discussion concerning the future of Grünau is still going on. The *Development Strategy Grünau 2020* is a necessary update in light of current frame conditions and forecasts (Stadt Leipzig 2007).

Regional Development) (Rössler et al. 2005).

4 For this purpose, different methods of qualitative and quantitative research were used. An important part of the research project was the questionnaire survey used in the context of the study “Dwelling and Living in Leipzig-Grünau” (Kabisch et al. 2005; Kahl 2003).

5 The Urban Development Concept contains different sections. Here, the part dealing with the large industrial housing stock of the whole city of Leipzig is meant, which includes the Grünau housing estate.

Results of the study

The study revealed that the success of the urban restructuring approach intended to reinvigorate large housing estates with green spaces strongly depends on ideas related to the design and function of the sites in question. Furthermore, this approach is particularly determined by financial, legal and organisational conditions. In Grünau, the open areas resulting from the demolition of buildings are predominantly designed using low-priced measures and simple equipment. This has resulted in less useful green spaces with high maintenance costs for the owners. Despite financial restrictions and anticipated oversupply of public green spaces, few new types of green spaces are being realised. However, some sites are designed with new demands and innovative ideas in mind. There are several examples of temporary design, areas with private responsibility and allotments for rent (Fig. 2). The study also detected many restrictions on the sustainable development of green spaces. On one hand, there were failures in the strategic planning of the municipal authorities regarding the integration of green space issues. On the other hand, there was a lack of ideas regarding the design and the function of the increasing number of green spaces that dealt with changing needs and aesthetic values. Owners, local administration and residents have different interests, expectations, opportunities and restrictions within the process of urban restructuring, all of which influence the scope of actions. The main problems and recommendations are described below.

The role of the municipality

Overall strategies and visions for Grünau are obviously based on ideas for green space development; a fact that emphasises the importance and relevance of the issue. This also emphasises the challenges resulting from the understanding that green spaces represent a solution to structural problems. However, in contrast to these ambitious objectives, there is currently a lack of integration of green space objectives into planning and implementation. There was no verified green space strategy or concept in place at the start of the process. Basic studies concerning ecological and green space planning issues do exist but, for the most part, they have not been included in urban planning documents. Partial green space planning still suffers from a relatively low level of importance and from being subordinated due to economic demands. The primary results of this have been a lack of integration of the environmental and green space planning departments of the city into the restructuring process. Therefore, green space issues have not formed a part of the city's thinking concerning the restructuring of the urban fabric. Comparing the objectives of the *Urban Development Concept* with the current situation, the concept does not appear to have been a strong instrument for representing municipal interests. The barriers for implementation are mainly caused by the differing interests of the municipal administration, owners and residents.

Fig. 5: Empty former building site – simply lawn

Fig. 6: Active participation of residents in the realisation of new green spaces

Indeed, the most difficult task in urban restructuring processes is connecting town planning objectives with aspirations of owners and the housing market. In light of the city's ambitious planning objectives, the demolition of properties by their owners may lead to conflicts and strains caused by the two different and competing aspirations. The reality in Grünau shows that often decisions taken to demolish certain buildings were made independently of municipal strategies and plans but with the clear support of the owners. The results are more or less dispersed areas of demolition and difficulties in realising overall concepts, particularly for green space planning (Fig. 5). Furthermore, local authorities and the city administration have neglected formulating clear statements for a future strategy that deals with the current demographic, social and structural problems of Grünau. Thus, for both owners and stakeholders alike, the existing concepts do not lead to reliable statements related to planning issues and the projected future of Grünau.

The current discussion about the *Development Strategy Grünau 2020* shows increasing awareness of future requirements for a successful restructuring process. The necessity of demolishing certain structures so that the remaining housing stock is spatially concentrated has been recognised. Thus, it appears possible that the future housing stock ought to be stabilised by the strategy, the further scattering of green and open spaces can be avoided and that establishment of advantageous open and green spaces with regard to their distribution and dimension can be fostered (Stadt Leipzig 2007:6). Unlike the first *Urban Development Concept*, the current *Development Strategy Grünau 2020* defines areas, following the urban model of "Duality of a Core and a Restructuring Belt" (Stadt Leipzig 2007:8):

(1) Where demolition activities should be concentrated and
(2) Where different measures for revaluation aiming at the long-term stabilisation of the housing stock and neighbourhoods should be realised.

Seen from the perspective of spatial development and green space issues, this concept is advantageous in many respects, even if there are strong reservations on behalf of residents and, to a certain extent, on behalf of owners. The necessity of a comprehensive concept or strategy for green space development has also been recognised (Stadt Leipzig 2007:7; annex:3) – although such a concept has been missing until now.

The constraints of the owners

Through implementation of the government programme, building owners have become the central actors in the realisation of measures meant to promote urban restructuring (for the relationship of the municipality and the building owners see also the paper of M. Bernt in this book). At least, the larger housing associations, for the most part, have been making their decisions concerning which buildings are to be demolished in accordance with their own economic criteria and often independently of administrative demands. The behaviour and decisions of smaller owners in Grünau, however, are characterised by various forms of uncertainty regarding the future of the housing estate and investments caused by the lack of clear and binding statements of local authorities, the ongoing population loss and the negative image of the neighbourhood. In general, real estate owners are aware of the fact that green spaces are important and that they are quality-determining elements of the residential environment. Nevertheless, there is little willingness to invest in green spaces. The mixed structure of ownership complicates the clear reference of improved green spaces to the residential buildings of the same owners.[6] Owners are principally interested in solutions that cost less yet have positive economic effects on the adjoining housing stock (e.g. parking spaces and allotments). In terms of governmental aid, grants to design sites of former buildings can be paid but owners have difficulties ensuring continuous maintenance of these spaces to preserve their quality and usability and therefore the sustainability of public investments.

The demands of the residents

The results of a questionnaire survey in Grünau emphasise the significance of green spaces to residents' satisfaction with their environment. Fig. 7 shows the level of satisfaction with different facilities of the residential environment – the satisfaction with green spaces was more critically rated than issues of retail and traffic infrastruc-

6 During the last few years, a very heterogeneous structure, with 18 private companies being owners of the housing stock, has been established (Bernt 2005).

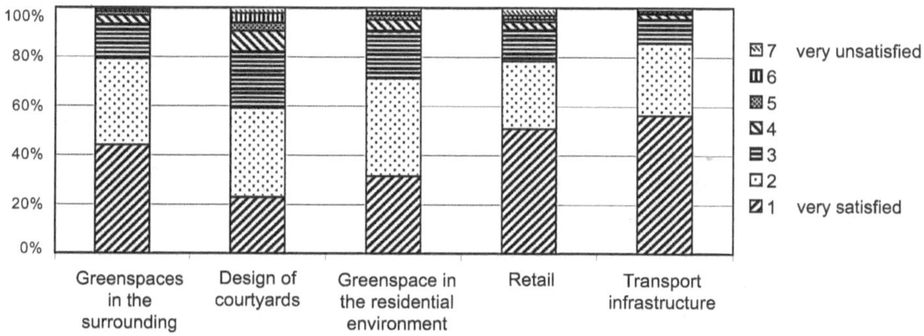

Fig. 7: Satisfaction with Grünau (STEFANIE RÖSSLER, data questionnaire survey UFZ 2004)

ture. The provision of green spaces is satisfying from a quantitative point of view. However, there is a lack of multifunctional spaces and provisions for different age groups. In general, the green environment and potential of open spaces in Grünau are understood as being an important benchmark for further development. For this reason, residents critically assess new approaches and deficits in green space planning, equipment and maintenance.

Reduction of maintenance, instigation of more natural processes and enhancement of self-determined activities for residents are the logical consequences of financial restrictions (Fig. 6). However, it appears that residents need more time to get acquainted with and accept the new types of green spaces. Most owners and responsible local departments have recognised the importance of involving residents in order to determine their needs, ideas and concerns regarding green space development. Due to social problems, migration of residents, uncertainty about development perspectives and significant past disappointments there is a low level of continuous and constructive engagement of a large proportion of the local population.

4. Conclusions and recommendations

Awareness of green space issues within the restructuring process at Grünau is relatively high. Nevertheless, the perception of local authorities and owners is quite different, especially regarding financing, collaboration and the shape and function of green spaces. Besides this, in the context of the current restructuring process, insufficient attention is paid to the demands and requirements of green space planning. Therefore, the chance to transform the neighbourhood with the aim of promoting a better quality of life and environment is being neglected. The process is also negatively affected by the lack of green space strategies and overall concepts. Ecological desires and the need to develop green spaces play a subordinate role in the process. Furthermore, existing ambitious objectives of urban restructuring activities and green space visions to improve the quality of life are not followed up by adequate integration of green space issues into planning documents, decision-taking and implementation. Thus, in the context of the ongoing urban restructuring process, there has been a declining relevance and impact of green space and ecological development.

To achieve more proactive green space planning it is necessary (Rössler 2003):
(1) To define planning guidelines and clear strategies,
(2) To develop strategic and independent concepts for green space development,
(3) To include basic documents concerning ecological demands in concepts and master plans for restructuring,
(4) To integrate green space planning into master plans on a city and district level,
(5) To foster the use of informal instruments
(6) To point the economic value of green spaces as location factor
(7) To create new funding methods and
(8) To involve all stakeholders equally in decisions relating to the process of urban restructuring.

The criteria employed for demolishing buildings should be reviewed in order to strengthen the requirements of green space planning. To ensure a high quality of development, the reuse of cleared estates should be considered before any demolition work occurs. Clearly defined minimum standards for design that work within financial constraints should be defined, in order to support the quality of life in the neighbourhoods in question. Successful future green space development needs innovative types of urban green spaces, new kinds of financing and management and alternative concepts for maintenance. To realise these demands it is necessary to integrate residents and future users into the planning and implementation of urban restructuring processes. Although owners benefit from governmental programmes to revitalise large housing estates, they are not adequately held responsible for the reuse and the maintenance of demolition sites. Especially because of granting financial aid local authorities should have more influence and instruments to manage restructuring processes to ensure sustainable development and investments.

References

BBR (2004): Zwischennutzungen und neue Freiflächen. Städtische Lebensräume der Zukunft. Berlin.

Becker, C. W. (2003): Sparen mit Konzept. Garten + Landschaft 113 (11): 26-28.

BERNT, M. (2005): Die politische Steuerung des Stadtumbaus in Leipzig-Grünau. UFZ-Diskussionspapiere (24). Leipzig.

BERNT, M. & S. KABISCH (2006): Ostdeutsche Grosswohnsiedlungen zwischen Stabilisierung und Niedergang. DISP 42 (164): 5-15.

BLUME, T. (2005): New Urban Configurations In: IBA Office: The other cities. IBA Stadtumbau 2010. Band 1: Experiment. Dessau: 111-116.

BMVBS & BBR (2007): Zweiter Statusbericht. 5 Jahre Stadtumbau Ost – eine Zwischenbilanz. Berlin.

BMVBW (2005): Nachhaltige Stadtentwicklung – ein Gemeinschaftswerk. Städtebaulicher Bericht der Bundesregierung 2004. Berlin.

DOEHLER-BEHZADI, M., KLEMME, D. A., KOCH, M., LÜTKE-DALDRUP, E., REUTHER, I. & K. SELLE (2005): Planloses Schrumpfen? Steuerungskonzepte für widersprüchliche Stadtentwicklungen. Verständigungsversuche zum Wandel der Planung. DISP 41 (2): 71-78.

GISEKE, U. (2002): Versorgungsgrün – kein Leitbild für die perforierte Stadt In: KORNHARDT, D., PÜTZ, G. & T. SCHRÖDER: Mögliche Räume. Hamburg: 167-173.

HEILAND, S. (2005): Urbane Räume im Wandel. Anforderungen an Naturschutz und Landschaftsplanung. Naturschutz und Landschaftsplanung 37 (1): 21-28.

HESSE, M. & S. SCHMITZ (1998): Stadtentwicklung im Zeichen von „Auflösung" und Nachhaltigkeit. Informationen zur Raumentwicklung 1998 (7/8): 435-453.

HUTTER, G., WESTPHAL, C., SIEDENTOP, S., MÜLLER, B. & G. JANSSEN (2004): Handlungsansätze zur Berücksichtigung der Umwelt-, Aufenthalts- und Lebensqualität im Rahmen der Innenentwicklung von Städten und Gemeinden – Fallstudien. UBA-Texte (41). Berlin.

KABISCH, S., BERNT, M. & A. FRITZSCHE (2005): Grünau 2004. Einwohnerbefragung im Rahmen der Intervallstudie „Wohnen und Leben in Leipzig-Grünau". UFZ – Umweltforschungszentrum Leipzig-Halle GmbH. Leipzig.

KAHL, A. (2003): Erlebnis Plattenbau – Eine Langzeitstudie. Reihe Stadtforschung aktuell, Band 84. Opladen.

KÜHN, M. (2000): Vom Ring zum Netz? Siedlungsstrukturelle Modelle zum Verhältnis von Großstadt und Landschaft in der Stadtregion. DISP 36 (143): 18-25.

NOHL, W. (2002): Freiraumplanung zu Beginn des 21. Jahrhunderts: Gesellschaftliche Entwicklungen und ihr Einfluss. Stadt + Grün 51 (8): 9-16.

PREISLER-HOLL, L. (2004): Freiflächenmanagement. Aspekte der Finanzierung, der Sicherung von Qualitätsstandards und des Unterhalts. Informationen zur Raumentwicklung 2004 (11/12): 679-686.

REUTHER, I. (2002): Leitbilder für den Stadtumbau In: BMVBW & BBR: Fachdokumentation zum Wettbewerb „Stadtumbau Ost": Expertisen zu städtebaulichen und wohnungswirtschaftlichen Aspekten des Stadtumbaus in den neuen Ländern. Bonn: 12-24.

Rat für nachhaltige Entwicklung (2004): Mehr Wert für die Fläche: Das „Ziel-30-ha“ für die Nachhaltigkeit in Stadt und Land. texte (11). Berlin.

Rössler, S. (2003): Schrumpfung – neue Chancen für die Freiraumentwicklung? Forderungen an den Stadtumbau aus Sicht der Freiraumplanung. Stadt + Grün 52 (11): 14-18.

Rössler, S., Bernt, M. & S. Kabisch (2005): Interessensgegensätze erfordern neue Umsetzungsstrategien. Freiraumentwicklung und Stadtumbau in der Großsiedlung Leipzig-Grünau. Stadt + Grün (9): 15-20.

Schmidt, A. (2005): Konsequenzen aus der demographischen Entwicklung für den Umbau der Grünflächen älterer Wohnsiedlungen und öffentlicher Parks. Raumforschung und Raumordnung 63 (3): 210-215.

Seitz, H. (2006): Nachhaltige kommunale Finanzpolitik und demographischer Wandel. In: Bertelsmann Stiftung: Wegweiser Demographischer Wandel 2020. Analysen und Handlungskonzepte für Städte und Gemeinden. Gütersloh: 180-186.

Stadt Leipzig (2002): Stadtentwicklungsplan Wohnungsbau und Stadterneuerung: Teilplan Großsiedlungen. Beiträge zur Stadtentwicklung (34). Leipzig.

Stadt Leipzig (2004): Zwischenbericht Grünau 2004. Kleinräumiges Monitoring der Stadtentwicklung in Leipzig. Leipzig Stadtplanungsamt.

Stadt Leipzig (2005): „Fallbeispiel Leipzig-Grünau“. Porträt im Rahmen des Projektes LHASA. Leipzig ASW.

Stadt Leipzig (2007): Entwicklungsstrategie Grünau 2020. Leipzig.

Tessin, W. (2004): Freiraumverhalten und Soziologie. Soziologische Aspekte der Nutzung und Planung städtischer Freiräume. Eine Einführung. Wiesbaden.

TRAMS AND THE REGENERATION OF BROWNFIELD SITES

André Mulder

Brownfield sites, such as those created by the relocation or disappearance of ports and factories, can be large-scale, unappealing areas within cities, but they also create the opportunity for regeneration. However, selection of new activities to fill these often large open spaces can be difficult, especially in shrinking cities. To make matters worse, brownfield sites often lack good transportation links even if they are close to the heart of a city.
Looking at the examples of Manchester, Sheffield and Amsterdam, this chapter reflects on the ways public transport, especially trams, can help to make regeneration a success.

1. Introduction

Many cities have lost the industrial activities that created them and generated growth for over a century. In other cities, ports and large industrial estates still exist, but have been relocated, usually further away from the city centre. In both cases, underused or even empty large areas are left. These places are often unappealing, although sometimes industrial 'monuments' give a certain atmosphere.
When industrial or trade functions disappear, new demands for accessibility are introduced. Links that are perfectly acceptable for freight may be of little use for passenger transport. Therefore transformation of an industrial or trade site to, for example, leisure or housing, can render its infrastructure inadequate. Although sometimes this can be adapted to new demand, like the freight-only railway line in Sheffield that was converted into a new passenger tram route.
The need to build new infrastructure can also be considered an asset; a positive move to attract general interest and investment. This approach was chosen in Manchester, where the new tram system was extended to Salford Quays, a dockland redevelopment project, with the specific aim of revitalising the area.

This chapter attempts to answer to the question: Does investment in public transport help to ensure successful urban regeneration? The redevelopment of brownfield sites will be taken as an example, sites in developed urban areas that have been affected by their former use (for instance as an industrial area), that are derelict and underused, and require intervention to bring them back to benefical use.[1] This interven-

1 This chapter is building on the definition of brownfield sites as developed by the CABERNET network (see chapter 2).

tion can take the shape of introducing new public transport links, and this is what this chapter is about.
The focus is on trams, as these are highly visible parts of the regeneration process and can easily be adapted to the specific requirements of sites (for example, low-speed through shopping centres and high-speed in low density areas). Three case studies, Sheffield, Manchester and Amsterdam, will be discussed. For each case, the reasons why the new infrastructure was built (or old infrastructure adapted), and the foremost participants and stakeholders will be identified. As all of these projects were finished quite recently, it is not possible to give a final evaluation, but is it is possible to assess their success to date.
However, this poses three interrelated questions concerning the methodology used in the case studies and the definition of concepts that are briefly addressed here:

What exactly is regeneration? For the purpose of this article the working definition developed by Bürkner et al (2005) is used.: Urban regeneration indicates demographic, social economic and physical renewal processes in cities and neighbourhoods that have been hit by problems concerning the loss of jobs and inhabitants and buildings becoming vacant or decaying. It includes attempts to attract a new resident population (demographic renewal), the creation of jobs (social economic renewal) and the improvement or renewal of the building stock (physical renewal).
When can regeneration considered a success and how can success or failure be measured? Dabinett et al. (1999), concentrating on urban regeneration effects, used five factors to evaluate the effects of the introduction of the Sheffield Supertram: image; property values; change of land use; business investment; and the accessibility of jobs. They used interviews, indices of residential and non-residential building prices, the number of planning applications, and employee records to compare the situations that existed before and after the opening of the Supertram system. Their findings will be summarised in section 3 of this paper.
Is it possible to attribute success or failure to single policy measures, like improving public transport? This is the most problematic question to answer. An improved public transport system is never implemented in isolation; other developments, events and situations, like the decision of a big company to open an appartment store or its headoffice in the area, will affect the success of a project. However, Babalik's planning framework to predict the success of urban rail system, which will be discussed in section 5, may provide some clarity for the issue.

The primary method for compiling information about the case studies is the study of literature and statistics. All of the case studies were visited by the author of this article in the autumn of 2006, mainly to verify information and take photographs.

2. Manchester Metrolink

The city of Manchester, in the North-West of England, has approximately 400,000 inhabitants, and is part of the conurbation of Greater Manchester, which has 2,500,000 inhabitants. Manchester is an example of a city that was shrinking until recently due to the demise of the traditional manufacturing industry and consequent population loss. However, this trend now appears to have reversed.

Manchester was the first English city to reintroduce trams in the 1990s. Starting in 1992, a new tramway system called Metrolink connects the suburbs of Altrincham and Bury via the city centre, replacing rundown suburban railways that terminated at two different mainline stations: Victoria and Piccadilly, both located at the edge of the city centre. The existing railways were modernised and some new stations were added, but more importantly a new section of track along city centre streets was introduced, improving accessibility from the suburbs. The tram project was initiated by the Greater Manchester Council, a regional elected body (now abolished). Funding was provided by central government (33 %), the Greater Manchester Council and the private sector.

As far as passenger numbers are concerned, Metrolink is a huge success. By 2001 the trams carried double the number of passengers the train services did in 1987, exceeding predictions and leading to overcrowding during rush hours (PTEG, 2005). Urban regeneration was not part of the strategy (indeed most of the route serves affluent suburbs), but Metrolink appears to be an effective attraction for Manchester and almost certainly contributed to an increasing number of visitors to the central shopping area.

A second route, to Eccles via Salford Quays, was added in 1999. In contrast to the first route, the Salford Quays section of the second one was specifically designed to promote regeneration of a derelict canal-side industrial area (Fig. 1). Six stops provide access to major residential and business developments in the area. This new route was promoted by the Association of Greater Manchester Authorities, a non-elected regional body without formal powers. Only 11 % of funding was provided by central government, 67 % by the private sector (mainly the tram service operator), and the remaining part by the Greater Manchester Passenger Transport Authority (GMPTA, a non-elected public transport planning authority, like the German Verkehrsverbund, but without some of its powers) and the European Regional Development Fund.

Although the new route has not attracted similar passenger numbers to the first route (some 3 million passengers a year for a 8.2 kilometre extension as compared to 16 million passengers for the original 31 kilometre route), the arrival of Metrolink is considered as one of the "*key events* [that] *have helped to spur renewed investment activity*" in the Salford Quays Area (Raco & Henderson, 2005). Indeed, when projects, events or individual dwellings in the area are advertised, Metrolink is usually

Fig. 1: Salford Quays, Manchester: Trams near the old harbour basin, with modern office blocks in the background

mentioned as an incentive. The local council[2] and the private sector[3] alike consider the tram system as an asset for the development of Salford Quays (Fig. 2).

Manchester now has a tram network of 39 kilometres. Although a shortfall of patronage and increasing project costs in other cities led the government to withdraw support for most light rail schemes in the past, including proposed Metrolink extensions, the success of the Manchester trams has now led to a reversal of this decision, making Manchester probably the only place in England where new tram routes will be built for the next five years. Like the first route, the new extensions will make extensive use of existing suburban railways (in this case mostly running through poorer suburbs and rundown industrial estates), with sections serving old town centres along a route that is yet to be approved by the government.

2 "*Metrolink arrived at Salford Quays in 1999 running along the line reserved in the late 1980s, and promoted as an integral part of the dock 9 developments at the Anchorage and Harbour City. Shortly afterwards, it was opened all the way to Eccles, creating a major step change in the accessibility of The Quays.*" (www.salford.gov.uk).

3 On their "Why Invest?" page, Quay Investments describe Salford Quays as "*Manchester's London Docklands*", with "*excellent transportation links*". "*Salford Quays is served by public transport – including the Metro Link that runs through the heart of the Quays taking you to Manchester City Centre in 12 minutes. – Salford Quays also benefits from good links to the M60 orbital motorway, particularly via the M62 motorway spur.*" (www.quayinvestments.com). Note that public transport is mentionned first.

Fig. 2: Salford Quays, Manchester: Building work near Metrolink tram stop

3. Sheffield Supertram

With approximately 500,000 inhabitants, Sheffield in the North-East of England is slightly bigger than Manchester, although the conurbation it is part of (South Yorkshire, 1,300,000 inhabitants) is smaller. Like Manchester, Sheffield was a shrinking city in the 1980s and 1990s, due to the decline of its steel industry, but recently the population started to increase.

The story of the Sheffield Supertram begins in 1976, when the first plans were developed by the South Yorkshire Metropolitan Council (now abolished). Construction work began only in 1991, and by 1995 the system was fully operational (Haywood, 1999). After some rearrangement of through services, the system can now be described as two main routes and a branch route. Both main routes start in the Hillsborough Area in the Northwest of the city, running via the city centre to the new town of Mosborough and the Meadowhall out of town shopping centre respectively. The Hillsborough and Mosborough sections predominantly use newly laid tracks in the streets, while the Meadowhall section is a converted freight-only railway line (Fig. 3). This route runs through the Lower Don Valley, which used to be the home of traditional manufacturing industry, most of which has long disappeared. As such, it is a good example of urban regeneration by adaptation of existing infrastructure to transport passengers instead of freight. Central government paid for 91 % of the construction costs, the European Regional Development Fund 5.5 % and a small amount was contributed by the private sector (3.5 %).

In contrast to the Manchester scheme, the Sheffield Supertram never lived up to expectations. Annual patronage was estimated at 22 million passenger journeys, but during the first year the system was fully operational this was only 6.6 million. After lowering fares, improving main line frequencies (at the expense of a branch line) and introducing conductors on all trams, the number of passengers rose to 12.8 million in 2004/05 and is now expected to remain around that level.

Fig. 3: Sheffield: Meadowhall section of Supertram

Compared to the Salford Quays route in Manchester, the urban regeneration effect of the Sheffield tram system is considered to be insignificant. Indeed, there has been extensive research into the regeneration effects of the Supertram (see: DABINETT et al., 1999; HAYWOOD, 1999; LAWLESS & GORE, 1999). Dabinett et al. found that the introduction of trams increased the accessibility of jobs for people living around stops. However, Supertram had no measurable impact on the image of the areas it serves or on land use and business investment, while the impact on property values was negative.

The reasons for the lack of success of the Sheffield Supertram, both in passenger numbers and in encouraging regeneration, can be summarised as:

1. Planning Policy. The Sheffield Supertram did not play a key role in urban planning. Indeed, planning powers for the Lower Don Valley, where the Meadowhall section runs, were removed from the Sheffield municipality by the national government and handed over to the Sheffield Development Corporation (a non-elected body) in 1988. This corporation was the driving force behind a road widening scheme and demanded that developers included ample parking space. There was no attempt to concentrate early development around tram stops.
2. Road building. Furthermore, a major building program included road widening schemes in all corridors served by trams.
3. Routing details and spending cuts. Major new housing, office, and leisure developments were often located far from tram routes and stops. To avoid exceeding

Fig. 4: New leisure facilities near Sheffield tram stop

the budget, eleven out of thirteen planned Park & Ride (P&R) facilities were scrapped, although four of them were realised at a later stage, now making a total of six P&R facilities near tram stops.

4. Transport Policy. Bus services in England (outside London) were deregulated in the late 1980s, meaning that integration of bus and tram networks became unusual and outside the remit of local and regional authorities. Indeed, buses competed with trams (with lower fares and higher frequencies), and no feeder bus services to tram stops were opened until recently.
5. Housing Policy. Due to population decreases, and the existence of high rise estates with maintenance and social problems, some high density residential neighbourhoods were transformed into low density areas, at the same time as tram tracks were being built in adjacent streets.

It should be noted that this research was carried out during the first few years after completion of the network, and that passenger number have since risen. Indeed, the National Audit Office (2004) states that, in contrast to 1996, the Sheffield Supertram in 2003 delivered many of the expected benefits, including enhancement of the image of the city.

Other developments include the provision of new leisure facilities close to some of the tram stops on the Meadow Hall route in the Lower Don Valley (Fig. 4).

Expectations of the new Sheffield tram system were probably unrealistically high – a fact that proved to be detrimental to the tram system's image when it failed to deliver anticipated benefits. Indeed patronage per kilometre of tram route on the Sheffield system (12 million passengers boarding a year on a 29 km network) appears to be almost equal to that on the Manchester system (19 million passengers boarding a year on a 39 kilometre network), where expectations were lower. The *passengers boarding to route length index*[4] is 5.1 in Manchester compared to 4.4 in Sheffield. Therefore the Manchester system attracts 15 % more passengers as the Sheffield system. However, the *passenger kilometers to route length index*[5] tells a different story. This index is 53 for Manchester compared to 15 for Sheffield, so Manchester performs 3.5 times better. Comparing both indicators shows the difference between the two systems. Manchester Metro is mainly a high-speed suburban system, with few stops, while the Sheffield Supertram resembles the traditional urban tram systems of England, or many of the surviving European tram networks e.g. Germany or the Netherlands.

Plans to extend the Sheffield tramway system, including new routes to a hospital and Rotherham town centre, were recently rejected by central government because they are not considered to be cost-effective.

4. Amsterdam tram routes 10 and 26

In contrast to England, where all tram services (apart from Blackpool) had been abolished by the mid 1960s, the city of Amsterdam never abandoned its tram system. Trams still serve the city centre, inner city neighbourhoods and many new residential areas.
Originally, plans for the development of East Amsterdam included the building of a new underground railway line. This was later rejected, due to cost and a reduction of the number of houses and offices planned for IJburg and the eastern docklands respectively. It was then decided that trams would serve the eastern docklands, to be redeveloped into residential and mixed use neighbourhoods, and the new suburb of IJburg, a land reclaimation project. The short extension of tram route 10 to the eastern docklands opened in 2004, while trams started running on the new Central Station – Eastern Docklands – IJburg route 26 (8,5 km) in 2005. Building costs were € 23 and € 155 respectively, largely covered by central government grants. Building the IJburg route was part of an agreement between central, regional and local authorities, which followed the identification of major building locations by central

4 The number of passengers boarding (x 100.000) divided by the total length of the tramway system in km; Department for Transport (2006) statistics, author's calculation.

5 The number of passenger kilometers (x 100.000) divided by the total length of the tramway system in km; Department for Transport (2006) statistics, author's calculation.

government ('Vinex-locaties') in the Netherlands in 1993. Accessibility of the new residential areas and the wish to diminish car usage, especially during rush hours to prevent traffic jams, were the main reasons to build this route (www.ijtram.amsterdam.nl).

As both routes opened only recently, it is too early to comment extensively on passenger numbers or the success in helping regeneration of the eastern docklands. Nevertheless, after a slow start, during which technological problems leading to the cancellation of services were a dominant feature, passenger numbers on route 26 are now rising. Route 10 is expected to encourage commercial services to a 19th century neighbourhood that it passes on its way to the eastern docklands (Stadsdeel Amsterdam Centrum, 2002), but no research is being carried out to assess whether this is actually happening.

5. Discussion: When are trams succesful in helping regeration?

This chapter aims to answer the question: Does investment in public transport help to make urban regeneration a success?
BABALIK (2000) developed and tested a planning framework to predict the success of urban rail systems by means of a checklist. This checklist includes urban factors (such as the vitality of the city centre, and city form and density), socio-economic factors (the number of passengers and their income levels already using public transport in areas along the route), planning factors (e.g. planning of the rail system in conjunction with a metropolitan development plan), operating policies (like high frequencies and the introduction of travel cards) and transport planning policies (e.g. integration of bus and rail networks, the introduction of Park & Ride facilities and a restriction of car parking in the city centre). Testing this checklist, Babalik found that the negative or positive performance of a new urban rail system can to a large extent be explained by the items on the list. This indicates that it is possible to predict the success of a future urban rail scheme, and that it is also possible to develop a range of policies to make a (future) system perform more effectively.
For example, when a city centre is not economically healthy, policies may include city centre redevelopment projects, concentrating on areas near stations, and pedestrianisation. As laying tramway tracks involves engineering and road work anyway, projects like pedestrianisation of shopping streets can be implemented simultaneously, therefore minimising disturbance.
An example of socio-economic factors to be taken into account are the income levels in areas alongside proposed tram routes. If attracting passengers from upper middle

class communities is a priority, then quality of service would probably be the most important factor. If, on the other hand, customers using the service will mainly be people on low income, fare levels would be more relevant.
The National Audit Office (2004) collected data for the new English light rail systems and compared their performance with similar networks in France and Germany. Higher patronage levels in the latter countries can partly be explained by the subsidisation of fares, which in England is low to non-existing. French and German networks also tend to be larger and have more frequent stops reaching more customers, while routes were planned to connect with centres of social and economic activity.

Evidently investment in tramways can aid succesful urban regeneration. It can be a good way forward for brownfield sites that, according to the CABERNET definition, "*require intervention to bring them back to official use*", like the new tram route to Salford Quays near Manchester is showing. However, some projects, including the ones in Manchester and in many French cities, appear to perform better in this respect than others such as Sheffield. Indeed, as Haywood (1999) writes at the end of his paper about Sheffield: "*these conclusions underline the point made throughout the literature* [that] ... *transport can stimulate activity if planned to do so*". Or to put it a different way: introducing trams will only help urban regeneration if there is a planning process at work that (a) is intended to reach that goal; and (b) is implemented in a consistent and forceful way.

References

Babalik Sutcliffe, E. (2000): Urban rail systems – a planning framework to increase their success. Research paper: Summary of PHD-thesis.

Bürkner, H.-J., Kuder, T. & M. Kühn (2005): Regenerierung schrumpfender Städte – Theoretische Zugänge und Forschungsperspektiven. IRS Working Paper.

Dabinett, G., Gore, T., Haywood, R. & P. Lawless (1999): Transport investment and regeneration – Sheffield: 1992 - 1997. Transport Policy 6 (1999): 123-134.

Department of transport (2006): Light Rail Statistics – England: Key Facts.

Haywood, R. (1999): South Yorkshire Supertram – its property impacts and their implications for integrated land use-transportation planning. Planning Practice & Research 14 (3): 277-299.

Lawless, P. & T. Gore (1999): Urban regeneration and transport investment – a case study of Sheffield 1992-96. Urban Studies 36 (3): 527-545.

National Audit Office (2004): Improving public transport in England through light rail – report by the comptroller and auditor general. The Stationery Office House of Commons.

PTEG (2005): What light rail can do for cities – a review of the evidence. Passenger Transport Executive Group.

Raco, M. & S. R. Henderson (2005): From problem places to opportunity spaces – the practices of sustainable regeneration. Paper presented to the SUBR:IM Conference, March 2005.

Stadsdeel Amsterdam Centrum (2002): Raadsvoordracht inzake vaststelling eerste herziening bestemmingsplan Czaar Peterbuurt blok N41. Stadsdeelblad afd. 1 (73).

http://www.ijtram.amsterdam.nl (11-7-2006): Nut en noodzaak.

http://www.quayinvestments.com (9-7-2007): Why invest?

http://www.salford.gov.uk (9-7-2007): Metrolink comes to the Quays.

URBAN SHRINKING IN THE LIFE CYCLES OF CITY ARCHETYPES

Susan C. Wessman

1. Overview

The phenomenon of shrinking cities is not new; it recurs throughout history due to global and local, natural and human-generated causes. On a time-scale that is appropriate to cities, shrinking rarely results in any city's death, that is, in its actual disappearance from the face of the Earth. Like biological organisms or communities, cities have life cycles: they germinate, grow, propagate, and deteriorate. Unlike an organism, however, death does not inevitably follow deterioration. Cities are more like biological communities in that deterioration can lead to re-growth. For both urban and biological communities, deterioration takes many forms such as stagnation, shifting populations, and changing resources. Shrinking of a city's population and stagnation of neighborhoods are two of the most striking indicators of urban deterioration. World-wide these are considered causes for alarm. The fact that a city shrinks *does* indicate that something is out of balance, but this shrinking also creates opportunities for regeneration. In this way cities seem to follow a 'succession', but one that differs from ecological succession in that cities *intentionally* regenerate themselves, often evolving along new pathways. This is possible because a city is made of people and institutions that can mobilize to collectively address the causes of deterioration and to initiate healing, much like a body's ability to heal an injury.

2. Questions and hypothesis

Even though death is not inevitable when a city shrinks, clearly many cities have shrunk out of existence or down to a size that no longer qualifies as 'city'. Why do some cities persist through shrinkage and others not? What is different about cities that have persisted through cycles of growth and shrinkage as compared with those that have shrunk and not recovered? One way to approach this question is to look for patterns and cycles of urban development and decline over time in order to identify and understand underlying causes of growth and shrinkage. This might help us understand how survivor cities assembled and used resources to reinvent and regenerate themselves. My working hypothesis is that by understanding the behavior of these long-lived cities we likely can better understand our own cities of today. By the same token, failure to understand the historic trajectory that helped form the contemporary character of an individual modern city may limit our ability to envision solutions that use shrinking as an opportunity to evolve and regenerate.

3. Method

The urban patterns and life cycles studied and discussed here are the generally accepted historic city archetypes described by Lewis Mumford (see Suggested Readings) . A 'city archetype' is the idealized form and character of the family of cities that incorporated the prevailing urban form and purpose during a distinct historic phase. Each city archetype embodies the ideas and culture of its age.

The real cities of a particular archetype were (or are) of varying sizes, but all fit the definition of 'city' - a permanent settlement with a high concentration of people, under unified control, forming a highly differentiated and specialized community, and serving purposes that transcend mere survival. The city is now, and historically has been, where the widest possibilities for contact and conversation exist because of the variety and density of people and activities.

Urban shrinking is governed by internal and external agents. Internal agents include human-generated, or what we might call 'behavioral', conditions such as war or population changes. External agents are naturally-occurring events such as earthquakes or the silting up of harbors. Both internal and external agents can be global, simultaneously affecting many cities, or more local (one city by itself). This paper focuses on internal agents of shrinkage (global and local) because they are the most likely candidates for revealing trends caused by behavioral conditions.

To find those trends I considered 'shrinkage' in each archetype, taken in historical order from oldest to today. For each archetype I asked basic questions that could reveal the essential character of that archetype. I expected that patterns of behavior that contributed to its decline or longevity would emerge. The questions I considered for each archetype were:

1. Original purpose of the city
2. Dominant planning paradigm that determined city form
3. Initial power structure and social structure
4. Emergent phenomena that arose and changed factors 1-3 above
5. Cause of shrinkage (or death)

In addition to answering the questions above, the cycles of birth and death of archetypes were mapped along a timeline, using an archetype's appearance as a city type of power and importance to mark 'birth' and the decline or transformation of the city's power to mark 'death' (Fig. 1). The length of an archetype's bubble along Fig. 1's time axis represents how long it dominated city form and function. The bubble's transition from dark gray to white indicates the decline of that archetype as a new one is arising.

4. Results

For at least 5000 years leaders have founded cities at locations that had become permanent collections of activities. They did so to serve an essential purpose and to manifest an urban form that was based on guiding principles, beliefs, and needs. Once built, these original forms tended to grow in chaotic, decidedly opportunistic ways that modified and added to the original purpose and form. In this way the city became a place with a high concentration of people and activities. People were initially attracted to these early cities for the security they provided, but in time opportunities for trade and stimulation became just as important.

Because of the heightened opportunities for communication and exchange, cities became the major generators of new technology and knowledge. Cities were in turn shaped by those new technologies and systems of understanding. New technology and knowledge invariably led to a new urban type with a different form and purpose. Upon the evolution of a new city type, the old type either became obsolete or evolved into something different built upon (modifying) the older type. Some urban types achieved the status of 'archetype' because of their importance as transformers and generators of culture and their longevity as models for how cities should be designed and built.

Fig. 1 illustrates the major archetypes, their duration as the dominant urban paradigm, and the important wide-spread conditions in which each arose. It is important to see that periods of consolidation and urban growth alternated with long periods of chaos and urban disintegration.[1]

1 The timeline here is based on European history, but life cycle processes and urban archetypes can probably be generalized to urbanized and urbanizing regions throughout the world. For example, China had megalopolitan cities centuries before Rome became the first western example.

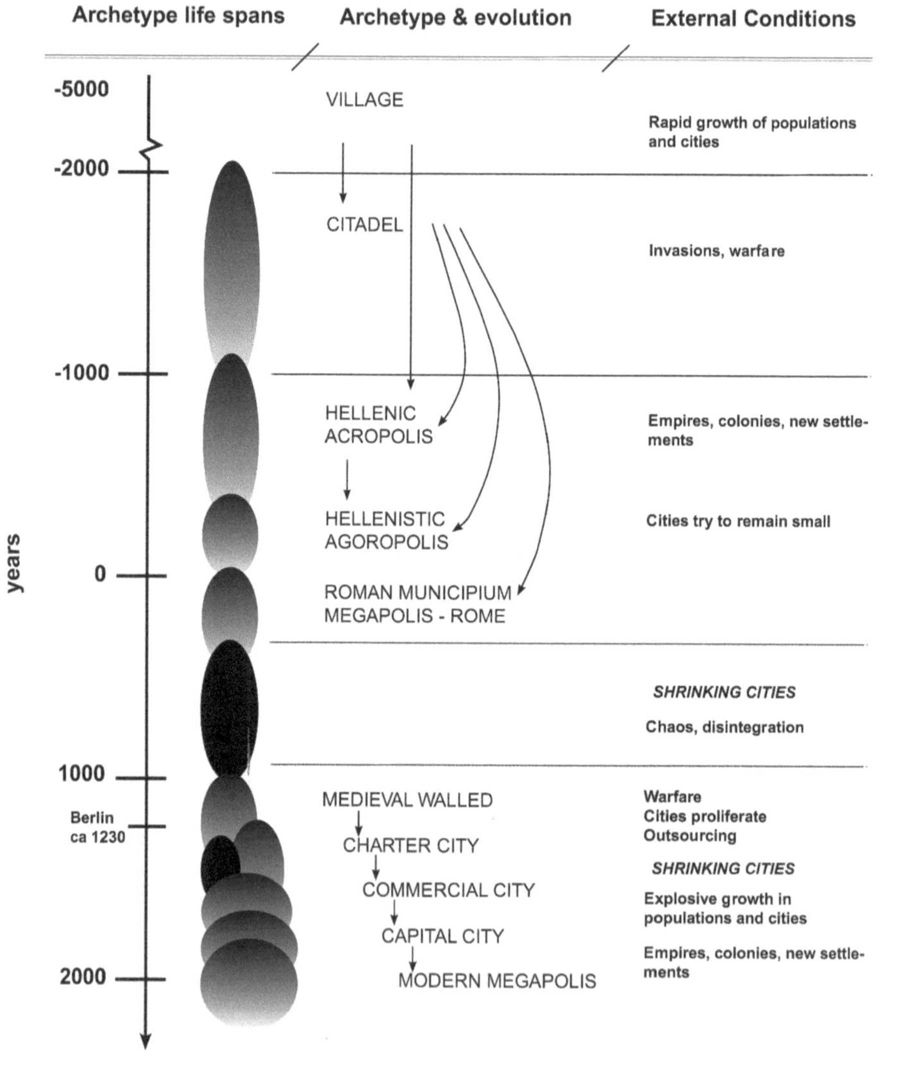

Fig. 1: Archetype life cycles and evolution trends

0-Village

Before the appearance of the first urban type at Uruk in the Fertile Crescent (about 7000 years ago), there was a period of rapid population growth and concentration of those populations in village clusters. This was an unstable time of warlord raids and attacks and the early urban form took shape as a walled, defensible fortress.

1-Citadel (Example: Uruk in Mesopotamia)

- Purpose: Enclosure, assembly. Stronghold for king, his gods and his goods. Security, protection for local villagers.
- Form: Double walled. Inner wall enclosed sacred precinct for the king's gods, palace, market, and granary. Outer wall enclosed collection of villages. Usually built around a well or spring.
- Power and social structures: Union of the sacred and the military; land and market belong to god(s) and priesthood (sacred), whose powers work through the king (military).
- Rural-urban relationship: Integral; villages and agricultural fields within the walled city.
- Emergent phenomena: Specialization and division of labor, and of precincts by function; standing military; agriculture; negotiated morality to resolve village/regional differences; land as a legal entity (Code of Hammurabi).
- Causes of local shrinking: susceptibility to physical conquest by invaders who wanted the city's wealth.

This simple walled rectangular form was used for thousands of years while warlords battled for control of trade routes and land. It persisted as the urban model until larger, denser urban forms were overlain or attached to the original walled form.

2-Hellenic Polis (Example: Socrates' Athens)

- Purpose: Trade and civics; development of intellectual culture.
- Power and social structures: Participatory democracy for citizen aristocracy. Culture and function of society based on slavery and on trade performed by non-citizens.
- Form: A 'union of villages' near a sacred site that was separate from government and trade. No protective walls, no palace (no king). Precincts: agora (gathering place for all, town hall, and market), park-arts-recreation, domestic services.
- Rural-urban relationship: strong and balanced; most residents still lived in villages, which surrounded the Polis.
- Emergent phenomena: replacement of king-god government by free-born citizenry; city planning standards (e.g., maximum population limit); imperialism to feed population.
- Causes of local shrinking: failure to create 'union of cities'; shift to tyranny and imperialism; power shift from aristocracy to wealthy merchants (non-citizens) who left to form their own cities elsewhere.

3-Hellenistic Metropolis (Example: Miletus)

- Purpose: Colonies spun off from too-large Hellenic Polis ('metro-polis' = mother city); movement of goods, armies; mercantile.
- Form: Grid pattern laid out as a unit; agora separated from town hall-government center and sited near port; application of Aristotelian/Platonic urban planning ideals.
- Power and social structures: Government as business run by professionals (bureaucracies); less participation of (or burden upon) citizens.
- Rural-urban relationship: Commerce and external trade more important than local agriculture.
- Emergent phenomena: First idealized, formalized city plan as an exercise in improving urban comfort (uniform street widths and block lengths, ease of movement); coined money; written alphabet.
- Causes of local shrinking: Attack from outsiders seeking city's wealth; rise of philosophies that reject materialism (after 500 bce).

4-Megalopolis (Example: Rome)

- Purpose: Governmental and economic seat of Roman Empire.
- Form: Accretions around ancient sacred precinct.
- Power and social structures: Predatory, empire building.
- Rural-urban relationship: Parasitic; Rome was too large to feed itself.
- Emergent phenomena: Vast, landscape-level planning coupled with spiritual beliefs; engineering (highways, aqueducts, architecture); universalized public entertainment.
- Causes of local shrinking: Overgrowth and consequent breakdown of urban infrastructure and external support structure; overreaching by the Empire; failure to create political and economic union of municipia and territories; repeated assaults and invasions.

5-Roman Municipium (Examples: London, Florence)

- Purpose: Outlying strongholds for Roman Empire; conduits for channeling the Empire's wealth back to Rome.
- Form: 'Centuriation': Identical planned layout applied to all cities; walled cities with precincts; all cities connected by paved roads.
- Power and social structures: All wealth and power accrued to Rome. Denizens were subjects to Rome, with no autonomy or voice in city operations.
- Rural-urban relationship: Unilateral (feed and supply Rome and her armies).
- Emergent phenomena: Street grid and cardo-decumanus framework: forum at central crossroads.

- Causes of local shrinking: Collapse of Rome. Some municipia continued to exist as lesser settlements if people chose to stay and live there.

First global cycle of city shrinkage: The collapse of the Roman Empire is the first instance of global-scale urban shrinking. It lasted hundreds of years (~400 to 1000 ce). Without the policing action of the Empire the urban infrastructure for most major cities in Europe and Asia Minor completely broke down, resulting in chaos.

6-Medieval Walled City (Examples: Hanseatic League cities, Venice)

- Purpose: Initial stages = survival, protection and security from continent-wide chaos and violence; regular market. Later stages = regional trade run by the guilds.
- Planning paradigm: Return to local, defensible, walled. Local ruler as 'wall-builder' to protect the market or monastery, which was the nucleus around which permanent settlements developed.
- Power and social structures: Evolved from king or priest to a balance between temporal (warrior, merchant, guildsman) and spiritual (priest, scholar); earned citizenship; military service a requirement of citizenship.
- Rural-urban relationship: Strongly bound together by local trade.
- Emergent phenomena: Clocks, bookkeeping, capitalism; large-scale provision for sick, old, poor; self-governance; improved transport by water; concept of functional urban zoning.
- Causes of local shrinking: Outsourcing (open market economies outside the city could out-compete closed market economy inside old city); inability of rigid walls and institutions to accommodate new ideas.

7-Charter City (Examples: Berlin ~1230; Salisbury)

- Purpose: Military strongholds and self-governance as established by 'charter of liberties' from king.
- Planning paradigm: Formulaic = wall and moat around the central town hall cum market; street grid with streets sized according to use; districts or ecincts based on function.
- Power and social structures: Legal rights and responsibilities for citizens; earned citizenship; king as giver of rights and collector of taxes.
- Rural-urban relationship: Unbalanced, with rural providing all necessary goods for urban survival, but without access to any of the privileges or rights possessed by urban citizens.
- Emergent phenomena: Land as source of income (rent) not agricultural produce; university, hospitals.

- Causes of local shrinking: Growth of international trade and mercantilism, which broke the guilds' monopolies.

Second global cycle of shrinkage. The second era of global-scale shrinking occurred in the Late Medieval period due to the confluence of predominantly human-generated causes: the plague, invasions, and crusades.

8-Commercial City (Examples: Berlin, London)

- Purpose: Unrestricted movement of goods; unlimited commercial enterprise.
- Planning paradigm: Infill, demolition, overlay of existing urban fabric; platting of rural areas (suburbanization) for development; emphasis on linearity to accommodate public and military transport.
- Power and social structures: Businessmen and entrepreneurs fill government positions; no sense of mutual responsibility between employers and employees.
- Rural-urban relationship: Rural areas as resource to be developed for and consumed by the city; urban as repository of and expression of wealth.
- Emergent phenomena: Fundamental change in historic city purpose (city had been 'for people', now 'for business'). Abstract market and speculation, privatization of feudal lands creates land as commodity; rapid public and commercial transport.
- Causes of local shrinking: Declines in market and supply; lack of well-positioned port.

9-Capital City (Examples: Berlin, Paris)

- Purpose: Permanent, centralized seat of governmental power.
- Planning paradigm: Influenced by physics (abstract, geometric, linear forms imposed on medieval city) and conspicuous consumption (large-scale planned destruction of neighborhoods to build palaces, boulevards, and new purpose-built districts).
- Power and social structures: Absolute power, despotism; highly organized; standing army.
- Rural-urban relationship: Rural as forced supplier of urban needs.
- Emergent phenomena: Home and workplace separate; change from goods economy to money economy; dominance of wheeled vehicles; bureaucracy; military engineering.
- Causes of local shrinking: Failure to also become a Commercial City.

10-Modern Megalopolis (Examples: Los Angeles, Mexico City, Berlin)

- Purpose: Acquisition of wealth; employment.
- Planning paradigm: Growth at any cost; vehicle-oriented; zoning.
- Power and social structures: Participatory (in varying degrees), hierarchical, bureaucratic.
- Rural-urban relationship: Mix of romanticized nature and rural and discomfort with having it too near, or indifference to/lack of awareness of rural and natural areas.
- Emergent phenomena: Instantaneous electronic communications; sprawl; annexation.
- Causes of local shrinking: Large shifts of population and commerce/business to the suburbs; decline of the urban core.

5. Intepretation

From the beginning, the primary purpose for founding cities was the attempt to transcend or dominate nature and to provide security. The driving forces behind the process are almost universally the drive of the city ruler to acquire wealth and power, and the citizenry's drive to gain employment and security. The city became a tool with which to amass, store, and protect that wealth, including cultural wealth. The city embodies power. This made cities the cause and source of much warfare throughout Western history.

For every archetype, technological innovation led to improvements in people's ability to acquire material goods and this led to de-stabilizing power shifts in the city. For example, in the Acropolis archetype (e.g., Socrates' Athens) trade expanded greatly. That led to a power shift from a small, elite group of citizens to a predominantly non-citizen merchant middle class that had little respect or need for the elite society that had excluded them from participating in running the city. The merchants thereupon founded cities of their own that competed with the Mother City: the Mother then shrank because it failed to adapt to an expanding, inclusive, market-based society.

There were two distinct periods of wide-spread, global-scale shrinking cities. Both phases were predominantly human-generated and occurred when cities were large and had relatively large populations at high densities. Shrinking of cities in the first period (about 400 to 1000 c.e.) was due to the collapse of the Roman Empire. Urban infrastructure completely broke down and people fled to the countryside as cities descended into lawless chaos and filth. In the space of a few decades Megalopolitan Rome shrank from one million inhabitants to 30-35,000. The second global phase of urban shrinking (about 1300 to 1500 c.e.) was due to the bubonic plague and the constant pressures of invasions and crusades.

Local scale and archetype-specific shrinking can be grouped into three categories: empire collapse, rigidity, and rapid growth. The first, or 'empire collapse' mode, is typical of the Hellenistic Agoropolis and Roman Municipium. Such cities depended on the support and infrastructure provided by an external power, because their primarily purpose was to act as strongholds for the empire and as inward conduits for its wealth. When their empires collapsed, such cities lost their infrastructure and security and either went into decline or found enough resources to re-invent themselves.

The second trend, rigidity, was typical of the Acropolis and Medieval archetypes. Such cities had rigid, specialized cultural institutions and physical forms that were unwilling or unable to adapt to change. Pressure to change came in many ways: different market and economy practices, rapid population growth that overwhelmed city institutions and structures (e.g., defense walls, sewers, water supplies), or shifts in rural-to-urban relationships. Failure to adapt usually resulted in wholesale breakdown of city institutions and form, and the founding of new cities elsewhere. This trend caused the first large-scale occurrence of 'outsourcing', because new enterprises could take root in rural areas where labor and land were cheaper and where there were fewer or no regulatory constraints.

The third trend, rapid growth, is typical of the more modern archetypes: Capital City, Commercial City, and Megalopolis. Such cities rapidly grew (and are growing yet) in wealth and size, and are distinguished by large-scale destruction of old neighborhoods to make way for large-scale construction of new quarters (e.g., for new palaces and grand boulevards in Capital Cities or commercial establishments in Commercial Cities). Capital Cities (even when purpose-built) that failed to evolve into commercial centers soon stagnated or shrank because power, prestige, and wealth moved to cities that *did* evolve into Commercial Cities. Capital Cities that became Commercial Cities in addition to remaining capitals survived to become the most powerful cities in their nations.

The climax this third trend is the Modern Megalopolis. Berlin follows this trend, beginning as a Corporate City (ca. 1230) then becoming a Capital City that evolved into a 'Capital + Commercial City' of immense power. Berlin attained megalopolis status in the first third of the 20th Century: in 1939 it was the sixth largest city in the world.

6. Discussion

Shrinkage is obviously not a modern urban phenomenon, nor is it irreversible, since many of our most beloved cities have been around for hundreds of years and have survived cycles of growth and shrinkage. In Europe, urbanism has weathered and recovered from two long-past but devastating 'global' eras of shrinking, plus many minor ones on more local scales. There have also been short periods of shrinking cities in the 20th Century, but at much smaller scales. For example, during the 1950s-60s in the U.S.A., businesses and middle-class families left city cores for the suburbs, leaving large portions of many cities unpopulated or populated only by people without the resources to maintain and protect those areas. Despite this shrinkage, none of those cities has yet died, or seems likely to do so. The real question is not "Expected date of death?" but rather "Into what form will each city change as it seeks to recover vitality and purpose?"

My original quest was to understand what reserves, attitudes, and planning are needed to reverse the decline experienced by a shrinking city. Two things stand out from this quick examination of the historical record:

1. Cities that have survived periods of local and global shrinking have consistently been able to withstand a period of disintegration and modification of their form and institutions. They were able to survive because they were populated by people who still wanted to live there despite the problems. The inhabitants identify with the city, it is symbolically and physically important to them. Many of the oldest surviving cities are those that were built on sacred sites in pre-historic times and these cities have come to embody the culture and history of a whole civilization. They remain always as 'culturally important' places to live. In fact, shrinking seldom kills off the old city; there is too much 'will to live'. The old character and form persist in the new city and its form and add to the hybrid 'new' city's resilience.

2. Survivor cities had the necessary leadership, technologies, and monetary and material resources to fix or rebuild the city after shrinking or lying dormant.

7. Thoughts on today's shrinking cities

To help a city that is undergoing deterioration and shrinking we must be able to answer these questions: (1) Is the shrinking is due to local or global causes? (2)

What are the city's historic archetypal patterns and growth trajectory? The answers will help us detect entrenched patterns of behavior and attitudes, identify conflicting paradigms, and see how to develop a balanced new approach. It is crucial to see how the assumptions underlying currently dominant planning paradigms affect our ability to envision alternative solutions. In the United States the underlying paradigm is that growth of tax-base and revenue must be continual, which seems to fly in the face of physical and ecological realities. That paradigm very seriously constricts the available options for meeting changes. We should keep challenging such basic (often hidden) assumptions and ask three very important, and often emotionally charged, questions:

Is shrinking necessarily bad?
As pointed out by the Shrinking Cities exhibit in Berlin (2004), vacancies created by shrinkage are opportunities for new and creative enterprises. These are the places where the seeds of new enterprises can take root if the process is not prevented by antiquated paradigms locking in a city's form and functions. Vacancies and momentary shrinkages happen all the time in ecosystems: after a forest fire or even the fall of a single huge tree, new seeds will take root and begin the succession process. Perhaps this is an urban succession pattern: a city begins to shrink, sections of the city die or lie fallow, artists and students discover the unwanted section and begin to invigorate and recreate it. In time, the section becomes attractive to a new class of people who move in and recreate it again.

Is constant growth necessarily good (or even possible)?
We are in the era of the Modern Megalopolis – cities large and small annex other areas and towns with the intent to grow the tax base and increase revenues. But at some point city traffic becomes overly burdensome, the impact on the surroundings irreversible, and the quality of life declines. A shrinking city is a signal that resources are out of balance and communication is breaking down somehow. From the definition of 'city' given earlier, an essential part of how a city functions is the quality of its internal communications: how well do different functions, agencies, communities understand and communicate with each other? The Megalopolis archetype may be a unique urban form because it is more like an accretion of independent, barely-communicating institutions and districts, rather than an assembly of inter-dependent functions and neighborhoods.

How big is big enough?
At what size does a city with a healthy diversity of urban peoples and activities (desirable characteristics of a city) become too large and break into fragments of smaller, more manageable sizes? If one of the defining characteristics of 'city-ness'

is increased possibilities for communications and interactions, then there may well be a point at which there is an overload and face-to-face exchanges become something to avoid. This reflects back to the quality of urban intra-communications.

Berlin: 'survivor city'?

Berlin, a relatively 'young' European city, is over 700 years old and has experienced many cycles of local and global shrinking and stagnation. Among the most important considerations for Berlin, as it faces another cycle of shrinkage, are: (1) to be sure that all the operating paradigms of urban form and function are recognized, because they will invisibly and strongly shape the conversation; (2) to use the conversation about paradigms as a tool for developing good communication among neighborhoods and among governmental groups; and (3) to avoid letting outmoded paradigms get in the way of ideas for multi-faceted revivals.

Acknowledgements

Many thanks go to my colleagues in the Urban Ecology Program at the University of Washington, Seattle, and the Graduiertenkolleg (RTG 780: Urban Ecology) of Humboldt University, Berlin, for inspiring discussions and the opportunity to present some of these ideas in Berlin in 2006. Special thanks go to Dr. Eric Shulenberger and Dana Wilk for discussing these ideas and this paper at length with me. I thank my employer Parametrix (Seattle, WA) and the UW Urban Ecology Program (funded by the US National Science Foundation) for financial support for my travels to Berlin. Finally, thank you to Lewis Mumford for his rich compendium, which inspired this too-brief summary.

Suggested readings

Bacon, E. N. (1980): Design of Cities. Middlesex.
Girouard, M. (1985): Cities and People, A Social and Architectural History. New Haven.
Griffe, M. (2003): Chronology of Asia Minor and Mesopotamia. Le Cannet.
Jacobs, J. (1961; 1992 edition): The Death and Life of Great American Cities. New York.
Mumford, L. (1961): The City in History. New York.

LIST OF AUTHORS

Uwe Altrock
Institute for Urban Development, Department of Urban Regeneration and Planning, University of Kassel
altrock@asl.uni-kassel.de

Katharina Appel
Formerly at Department of Soil Protection, Technische Universität Berlin

Matthias Bernt
Department Regeneration of Cities, Leibniz Institute for Regional Development and Structural Planning (IRS)
berntm@irs-net.de

Tobias Chilla
Institut für Geographie, Universität Erlangen-Nürnberg
tobias.chilla@fau.de

Valeska Contardo-Jara
Department Ecological Impact Research and Ecotoxicology,
Technische Universität Berlin
valeska.contardo@mailbox.tu-berlin.de

Ellen Diermayer
Formerly at Department of Geography, Humboldt-Universität zu Berlin

Wilfried Endlicher
Department of Geography, Humboldt-Universität zu Berlin
wilfried.endlicher@geo.hu-berlin.de

Miriam Fritsche
Bremer Institut für Soziale Arbeit und Entwicklung (BISA+E) e.V.
fritsche@bisa-bremen.de

Markus Hesse
Geography and Spatial Planning Research Centre, University of Luxembourg
markus.hesse@uni.lu

Mathias Hofmann
Zentrum für Interdisziplinäre Technikforschung, Technische Universität Dresden
mathias.hofmann@tu-dresden.de

Patrick Hostert
Department of Geography, Humboldt-Universität zu Berlin
patrick.hostert@geo.hu-berlin.de

Hadia Köhler
Formerly at Department of Geography, Humboldt-Universität zu Berlin

Ingo Kowarik
Department of Ecology, Technische Universität Berlin
kowarik@tu-berlin.de

Christoph Kulemeyer
Formerly at Museum für Naturkunde, Humboldt-Universität zu Berlin

Elmar Kulke
Department of Geography, Humboldt-Universität zu Berlin
elmar.kulke@geo.hu-berlin.de

Marcel Langner
Department of Geography, Humboldt-Universität zu Berlin
marcel.langner@geo.hu-berlin.de

Rudolf Maier
Department of Ecophysiology and Functional Anatomy of Plants,
University of Vienna
rudolf.maier@univie.ac.at

Peter Meffert
Formerly at Department of Ecology, Technische Universität Berlin

Harald Mieg
Department of Geography, Humboldt-Universität zu Berlin
harald.mieg@geo.hu-berlin.de

André Mulder
Department of Real Estate & Housing, Delft University of Technology
a.mulder@tudelft.nl

Gunnar Nützmann
Leibniz-Institute of Freshwater Ecology and Inland Fisheries, Berlin
nuetzmann@igb-berlin.de

Stefanie Rößler
Leibniz Institute of Ecological Urban and Regional Development (IÖR), Dresden
s.roessler@ioer.de

Anke Ruckes
Formerly at Institute of Geographical Sciences, Freie Universität Berlin

Ingeborg Schinninger
Formerly at Institute of Plant Science, ETH Zurich

Daniela Schüler
Formerly at Department of Geography, Humboldt-Universität zu Berlin

Marlies Schulz
Department of Geography, Humboldt-Universität zu Berlin
marlies.schulz@geo.hu-berlin.de

Elke van der Meer
Department of Psychology, Humboldt-Universität zu Berlin
vdmeer@rz.hu-berlin.de

Peter Werner
Institut Wohnen und Umwelt, Darmstadt
p.werner@iwu.de

Gerd Wessolek
Department of Soil Protection, Technische Universität Berlin
gerd.wessolek@tu-berlin.de

Susan Wessmann
Parametrix, Bellevue, Washington, U.S.A.
scwessman@msn.com

Janneke Westermann
Formerly at Department of Ecology, Technische Universität Berlin

Christiane Westphal
Formerly at Leibniz Institute of Ecological Urban and Regional Development (IÖR), Dresden

Claudia Wiegand
Department of Biology, University of Southern Denmark
wiegand@biology.sdu.dk

Betka Zakirova
Formerly at Institute of Geographical Sciences, Freie Universität Berlin

RESEARCH PROGRAMME
URBAN ECOLOGY
BERLIN

FSC
www.fsc.org
MIX
Papier | Fördert gute Waldnutzung
FSC® C083411

Kolibri 360 GmbH
Ferdinand-Jühlke-Straße 7
99095 Erfurt, Deutschland
produktsicherheit@kolibri360.de